Soil Science Education:
Philosophy and Perspectives

Soil Science Education: Philosophy and Perspectives

Proceedings of a symposium sponsored by Divisions S-1, S-2, S-3, S-4, S-5, S-6, S-7, S-8, S-9 of the Soil Science Society of America in Minneapolis, Minnesota, 5 Nov. 1992.

Editors
Philippe Baveye, Walter J. Farmer, and Terry J. Logan

Organizing Committee
Philippe Baveye, chair
Walter J. Farmer

Editor-in-Chief SSSA
R. J. Luxmoore

Managing Editor
David M. Kral

Associate Editor
Marian K. Viney

SSSA Special Publication Number 37

**Soil Science Society of America, Inc.
Madison, Wisconsin, USA**

1994

Cover Design: Patricia Scullion

Copyright © 1994 by the Soil Science Society of America, Inc.

ALL RIGHTS RESERVED UNDER THE U.S. COPYRIGHT ACT OF 1976 (P.L. 94-553)

Any and all uses beyond the limitations of the "fair use" provision of the law require written permission from the publisher(s) and/or the author(s); not applicable to contributions prepared by officers or employees of the U.S. Government as part of their official duties.

Soil Science Society of America, Inc.
677 South Segoe Road, Madison, WI 53711 USA

Library of Congress Cataloging-in-Publication Data

Soil science education : philosophy and perspectives : proceedings of a symposium sponsored by Divisions S-1, S-2, S-3, S-4, S-5, S-6, S-7, S-8, S-9 of the Soil Science Society of America in Minneapolis, Minnesota, November 5, 1992 / editor, Philippe Baveye, Walter J. Farmer, and Terry J. Logan ; organizing committee, Philippe Baveye, chair, Walter J. Farmer ; editor-in-chief, SSSA, R.J. Luxmoore.
    p.   cm. — (SSSA special publication : no. 37)
  Includes bibliographical references.
  ISBN 0-89118-809-6
  1. Soil science—Study and teaching—United States—Congresses. 2. Soil science—Study and teaching—Congresses. I. Baveye, P. (Philippe) II. Farmer, Walter J. III. Logan, Terry J. IV. Soil Science Society of America. Divisoin S-1. V. Series.
S591.55.U6S65   1994
631.4'071—dc20   94-17681
                                                        CIP

Printed in the United States of America

CONTENTS

	Page
Foreword	vii
Preface	ix
Contributors	xi

1 Introducing Soil Science into the K–12 Curriculum
    Terence H. Cooper, John W. Schultz,
    and Marion K. Barton 1

2 Undergraduate Core Curriculum in Soil Science
    K. A. Barbarick 7

3 Trends in Soil Science Teaching Programs
    J. Letey 15

4 Revision and Rescue of an Undergraduate Soil Science Program
    Ronald D. Taskey 21

5 Understanding Cognitive Styles: How to Teach to the Whole Soil Science Classroom
    Diana B. Friedman and Rodney J. Parrott 29

6 Private Sector Experience of a Soil Science Graduate
    Frances A. Reese 45

7 Advising M.S. Graduate Students: Issues and Perspectives
    Donald L. Sparks 51

8 Supervision of Ph.D. Level Soil Science Graduate Students
    Marion L. Jackson 59

9 Advising Doctoral Students in Soil Science
    Samuel J. Traina 65

10 The Advisor-Advisee Relationship in Soil Science Graduate Education: Survey and Analysis
    Philippe Baveye and Françoise Vermeylen 73

11 Educational Needs in Soils and Crops of Graduate Students from Developing Countries
    W. E. Larson, R. Kent Crookston, and H. H. Cheng ... 85

12 Advising Students from Developing Countries
   Elemer Bornemisza 99

13 Nontraditional Students: Off-Campus M.S. Degree
   in Agronomy
   W. L. Banwart and D. A. Miller 109

14 Distance Education in Soil Science: Reaching
   the Nontraditional Student
   Angelique L. E. Lansu, Wilfried, P. M. F. Ivens,
   and Hans G. K. Hummel 121

15 Fostering Learner Self-Direction in Soil Science Graduate
   Courses: A New Paradigm
   Philippe Baveye 135

FOREWORD

Two-thirds of Americans believe that science will improve their future and three-fourths indicate they enjoy learning science. Scientific literacy in this country, however, is at its lowest ebb since reaching a peak in the Sputnik era. The nation is currently undergoing a science and engineering reform with massive influx of public and private monies to counter this educational void. The goal is to attain preeminence in scientific literacy by the year 2000, perhaps a little presumptuous and unattainable given the timeframe. It is within this backdrop, however, that the relevance of soil science education takes on new meaning. Understanding the near-surface Earth properties, processes, and functionality is essential to global habitat sustainability. Soil science provides the educational framework to integrate components of earth-science systems, to understand the causes and consequences of spatial variability, and view dynamic processes impacting ecosystems in a holistic perspective. The future of our discipilne is heavily dependcent on our ability as educators and scientists to effectivley communicate this message. Our clientele are diverse. They represent multiple occupations, backgrounds, value judgements, interests, and experiences. Their understanding of soil and land resources may be limited. This special publication examines these issues, challenges, and opportunities for new trends in educational environments under new soil science paradigms. The Soil Science Society of America is committed to enhancing the outreach of earth-science education through development of resource learning materials and teacher mentor programs in conjunciton with the American Geological Institute textbook initiatives. We commend the authors of this text and the organizing committee for their efforts to bring this special publication to fruition in such a timely manner. It is a significant new contribution to the arsenal of earth-science educational materials.

LARRY P. WILDING, *President*
Soil Science Society of Agronomy

PREFACE

Traditionally, the "clients" of soil science education have belonged to three groups: undergraduate students, graduate students, and those clients reached through extension activities. In all three cases, the preferred mode of transmission of knowledge has usually been via formal lectures in classroom settings. Realizing the importance of the task before them, soil science educators have over the years paid significant attention to teaching in all of its multiple facets. In November 1969, for example, the Soil Science Society of America held a symposium devoted entirely to graduate instruction. During this symposium (and in its proceedings, published in 1970 as ASA Special Publication No. 17), prominent scientists were invited to analyze in detail the teaching needs and the knowledge-transmission methodologies used in the various subdisciplines of soil science (soil physics, soil chemistry, etc.).

Since the late sixties, numerous articles dealing with the teaching of soil science courses have appeared in, e.g., the Journal of Agronomic Education (to become the Journal of Natural Resources and Life Sciences Education), and the Agricultural Education Magazine.

Even though the visual aids (videofilms and multimedia technologies) available to instructors are currently evolving at a phenomenal pace, much of this existing literature on the teaching of soil science courses remains eminently relevant and useful. Therefore, when in early 1992 the SSSA Committee S571 ("Training of Soil Scientists") decided that the time was ripe to devote another symposium to soil science education and planned it for the fall 1992 annual meetings, it was agreed that this symposium should try to explore new facets of the field, to map out new territory, and not simply rehash information on what to teach and how to make lectures lively and appealing.

A number of directions that this symposium should explore seemed clearly dictated by recent trends and events. The need for a shift of emphasis from agricultural to environmental soils-related issues, brought about in part by the pronounced decline of farming in North America and Europe in the last two decades, mandates drastic changes in the soil science curriculum. The rapid pace of technological advances and the imminence of "information superhighways" challenge the need for and the usefulness of an extension service in its current form. Furthermore, universities must prepare their students for a life of continuous learning.

Aside from these societal changes, the university environment in which soil science educators have traditionally operated has also evolved significantly in the last two decades. The student body has become more and more diverse in terms of age and gender, while at the same time more foreign students are attending U.S. and Canadian universities than ever before. The areas of interest and training of the students and their career objectives also have evolved. All of these trends have combined to modify the conditions

conditions under which soil science educators have to approach the advising of undergraduate or graduate students.

These and other current challenges facing soil science education are analyzed in detail in the various chapters of the present SSSA Special Publication. These chapters have been purposely left heterogeneous in style and depth of coverage; some contributors have chosen to briefly relate their personal experience in a journalistic style, while others carried out detailed surveys or extensive literature reviews. In all cases, the authors benefitted greatly from the careful and thoughtful comments of anonymous reviewers.

May 1993

PHILIPPE BAVEYE
Cornell University

WALTER J. FARMER
University of Florida, Riverside

TERRY J. LOGAN
Ohio State University

CONTRIBUTORS

W. L. Banwart	Department of Agronomy, University of Illinois, 1102 South Goodwin Avenue, Urbana, IL 61801.
K. A. Barbarick	Department of Agronomy, Colorado State University, Fort Collins, CO 80523.
Marion K. Barton	Department of Soil Science, Borlaug Hall, University of Minnesota, Twin Cities Campus Minneapolis, MN 55455.
Philippe Baveye	Department of Soil, Crop and Atmospheric Sciences, Bradfield Hall, Cornell University, Ithaca, NY 14853-1901.
Elemer Bornemisza	Centro de Investigaciones Agronómicas, Universidad de Costa Rica, Código Postal 2060, San José, Costa Rica.
H. H. Cheng	Department of Soil Science, Borlaug Hall, University of Minnesota, Twin Cities Campus, Minneapolis, MN 55455.
Terence H. Cooper	Department of Soil Science, Borlaug Hall, University of Minnesota, Twin Cities Campus, Minneapolis, MN 55455.
R. Kent Crookston	Department of Agronomy and Plant Genetics, University of Minnesota, Twin Cities Campus, Minneapolis, MN 55455.
Diana B. Friedman	Department of Agronomy and Range Sciences, University of California, Davis, CA 95616.
Hans G. K. Hummel	Centre for educational production, Open University of the Netherlands, P.O. Box 2960, 6401 DL Heerlen, The Netherlands.
Wilfried P. M. F. Ivens	Department of Natural Sciences, Open University of the Netherlands, P.O. Box 2960, 6401 DL Heerlen, The Netherlands.
Marion L. Jackson	Department of Soil Science, University of Wisconsin, 1525 Observatory Drive, Madison, WI 53706.
Angelique L. F. Lansu	Department of Natural Sciences, Open University of the Netherlands, P.O. Box 2960, 6401 DL Heerlen, The Netherlands.
W. E. Larson	Department of Soil Science, Borlaug Hall, University of Minnesota, St. Paul, MN 55108.
J. Letey	Department of Soil and Environmental Sciences, University of California, Riverside, CA 92521.
D. A. Miller	Department of Agronomy, University of Illinois, 1102 South Goodwin Avenue, Urbana, IL 61801.
Rodney J. Parrott	Office of Instructional Support, Day Hall, Cornell University, Ithaca, NY 14853.

Frances A. Reese	Larsen Engineers, 700 West Metro Park, Rochester, NY 14623.
John W. Schultz	Department of Soil Science, Borlaug Hall, University of Minnesota, Twin Cities Campus, Minneapolis, MN 55455.
Donald L. Sparks	Department of Plant and Soil Sciences, University of Delaware, Newark, DE 19717-1303.
Ronald D. Taskey	School of Agriculture, California Polytechnic State University, San Luis Obispo, CA 93407.
Samuel J. Traina	Department of Agronomy, Ohio State University, Columbus, OH 43210.
Françoise Vermeylen	Office of Statistical Computing, Savage Hall, Cornell University, Ithaca, NY 14853.

1 Introducing Soil Science into the K-12 Curriculum

Terence H. Cooper, John W. Schultz, and Marion K. Barton
University of Minnesota
St. Paul, Minnesota

ABSTRACT

A survey of 129 teachers during the spring of 1992 was conducted to determine their usage of soil science concepts. Teachers represented both urban and rural school districts and grades K-12. The experience of the teachers surveyed averaged 20 yr. In the survey teachers indicated that they were not using soil science concepts because they were not part of the curriculum. The greatest percentage of teachers using soils concepts were the rural K-6 teachers and the urban 7-9 teachers. In order to increase the number of teachers using soil science concepts, the concepts must be included in the curriculum and teachers must be trained. The American Association for the Advancement of Sciences Project 2061 has outlined agricultural concepts that high school graduates need to know. Many of these topics are soils related and the opportunity for incorporating them into curricula is now.

INTRODUCING SOIL SCIENCE INTO THE K-12 CURRICULUM

The number of students studying soil science in Agricultural Colleges declined during the 1980s along with the total agriculture enrollment; however, recent trends (Litzenberg et al., 1991) indicate increases in enrollment in agriculture have not included the traditional soil science major. In order to increase the interest of students to study soil science in college it has been suggested that scientists assist in developing K-12 curricula that deal with soils (Barnes, 1987). Our purpose is to report on a study to determine the current extent of the usage of soil science concepts by a sample of Minnesota's elementary and secondary schools and determine ways that more of these concepts can be included into K-12 curricula.

SURVEY OF K-12 TEACHERS

A survey was mailed in 1992 to 45 schools (22 rural and 23 urban) selected at random to represent both the urban and rural population of Minnesota.

Copyright © 1994 Soil Science Society of America, 677 S. Segoe Rd., Madison, WI 53711, USA. *Soil Science Education: Philosophy and Perspectives.* SSSA Special Publication no. 37.

Survey for Teachers Concerning Usage of Soil Science Concepts in the Classroom

Surveys are needed back by May 30, 1992

Directions: Please complete the questions to assist us in determining your use of soil science. This survey will assist the Soil Science Society of America and the University of Minnesota in developing materials you may be able to use in the future. Thank you for your help. Please return all the surveys to the science coordinator who has the return address envelope.

Soil: Soil is the loose surface of the earth, capable of supporting plant life. All life on the planet Earth is dependent on the soil. The variability of our soil has determined many human activities. The degradation of the soil resource can alter the living standards of a society. Some examples of where soil science concepts can be used include: physical sciences (concepts of density and porosity and pH), biological sciences (soil organisms and plant growth), environmental courses (soil application of hazardous waste), social science (reasons for the dust bowl), geography (population centers and soil resources for food production).

1. Name _____ School _____
   Class grade taught = _____ years of teaching _____
   Course taught = _____

2. Have you ever used soil science concepts in your teaching?
   Yes ____ (please complete questions 3, 4, 5, 6) No ____ (please complete question 7 and 8 on back).

3. Please give some examples of your usage of soil science concepts in the classroom.
   a. Unit title = _____
   b. Concepts used = _____

4. Did you prepare your own soil science materials or were they prepared by other, and if so whom? Please explain.

5. What are the major problems you encounter in transferring information about soils to the classroom setting?

6. What specifically do you need (besides time) to be able to use more soil science concepts in your teaching?

7. The main reason (or reasons) for not using soil science concepts in our classroom.
   (Please rank the statements in order of importance 1 = most important; 4 = least important).
   Ranking:
   ____ Not part of my formal education.
   ____ I have had a class that mentioned soils (or a soils class) but did not determine how best to incorporate soils into my lesson plans.
   ____ I felt a lack of suitable classroom exercises or materials would doom their success.
   ____ Other (please list) _____

8. Would you be willing to use soil science concepts in your classroom if they were available in a suitable format?
   Yes ____ No ____ Unsure ____
   Comment _____

Your participation in this survey will put your name on a mailing list for future information about using soil science concepts in the classroom. The Soil Science Research Team thanks you for your help: Dr. Terry Cooper, University of Minnesota; Marlene Barton, Osseo High School; and John Schultz, Hopkins N. Junior High.

Fig. 1-1. Survey form used for soil concepts inventory.

Copies of the survey form (Fig. 1-1) were sent to the science or social science curriculum coordinator. Coordinators were asked to pass the survey forms on to the teachers. Coordinators collected the completed forms and returned them in the envelope provided. Eighteen schools returned the forms (6 urban and 12 rural) for a total of 129 teachers responding (63 rural and 66 urban). Urban schools included five in the Twin Cities and one in Rochester. Rural schools included three from the forested, northern region and nine from the more agricultural, southern region of the state.

The teachers responding to the survey were experienced (Table 1-1) and were evenly divided among grades and subjects with the exception of the 10–12 science teachers in which only 14 responded.

The responses for the K–6 teachers for the question about the type of soil concepts used were similar for rural and urban teachers. The topic most often used was soil erosion. The main reason that soil concepts were not used was because they were not perceived to be part of the curriculum.

For Grades 7–9 teachers the responses were varied for the concepts used with the rural teachers using more agricultural related terms (fertilizer, lime, or soil pH) and the urban teachers using more environmentally related terms (water contamination or soil development). The main reasons for not using soils concepts was again the comment that they were not part of the curriculum and the lack of suitable classroom exercises. Even if teachers are interested in presenting a concept, if that concept is not a part of the current curriculum as determined by their school district, they most likely will omit it in favor of those that are in the curriculum.

The high school teachers used a number of soil concepts (N fixation or soil pH), with some differences indicated between the urban and rural groups (rural used fertilizer concepts). The main reasons for not using soil concepts were listed as time constraints and lack of suitable, prepared materials.

Table 1-1. Experience, subject matter and grades taught for teachers responding to soil science survey.

Grades taught	Subject area	Teachers	Years teaching		
			Mean	Standard deviation	Range
		no.		no.	
K–6	Social science and sciences	34	16.9	10.9	1–35
7–9	Social sciences (world history, geography, U.S. history, civics, or social science	31	20.5	9.4	1–32
7–9	Sciences (physical and earth science)	28	19.7	10.0	2–33
10–12	Social science (american, world and local history, and economics)	22	20.1	8.5	2–30
10–12	Sciences (physics, biology, and chemistry)	14	21.5	10.8	2–35

Table 1-2. Percentage of respondents using soils concepts: 129 teachers from 18 schools.

		Using soils concepts	Total teachers
		%	no.
K-6	Rural	41	
	Urban	15	34
7-9	Rural	14	
	Urban	30	59
10-12	Rural	11	
	Urban	14	36

Overall the greatest percentage of teachers using soils concepts were the rural K-6 teachers (Table 1-2) and the urban 7-9 teachers. It is noted that 30% of the social science teachers used soils concepts.

In summary, the main reasons for not using soils concepts were: (i) not part of the curriculum (grades K-6, 7-9, or 12-12), (ii) lack of current materials for classroom use (grades 7-9 and 10-12), (iii) lack of formal soils training (grades 7-9 and 10-12), (iv) not part of the textbook (grades 7-9), and (v) time constraints (grades 10-12).

PROJECT 2061

The American Association for the Advancement of Science has been involved in a project to improve science education in the schools (Bugliarello, 1989; Clark, 1989). Phase 1 of this project outlined agricultural concepts that high school graduates will need to know. Development of these concepts were performed by two panels: (i) physical and information sciences and engineering and (ii) biological and health sciences. Topics related to soils are included in the materials being developed and are currently being tested in selected schools. Examples of these include:

Physical Sciences

    the surface processes of the earth such as erosion, deposition, element cycles as well as the nature, distribution, creation, and destruction of soils;

    uses, availability, economics, politics, and dependence of society upon natural resources (soils),

Environmental Biology

    shaping of ecosystems, as it relates to soil nutrients and climate which determine the distribution and productivity of plants;

    in general, the total productivity of natural systems almost always exceeds the productivity of agricultural systems,

Human Ecology

    ecosystems will draw upon reserves of energy when available energy is insufficient, thereby depleting them, and at the same time unrecycled matter accumulates as pollutants;

American agriculture as a nonsustainable system that can be ameliorated by no-tillage farming;
failure to recycle causes additions of fertilizers that contaminate water supplies.

Soil scientists need to assist K-12 teachers to further refine how soil science concepts will be presented, especially if the concepts are controversial. Scientists also need to assist teachers in the development of materials that can be used in the classroom and encouraging them to add concepts into their curriculum.

NATIONAL SCIENCE TEACHERS ASSOCIATION

Dr. Gene Gennaro (1992, personal communication, College of Education, Univ. of Minnesota) has indicated that the National Science Teachers Association (NSTA) has a commitment to science-technology-society (STS) issues by developing a curriculum called *The Water Planet*. There is a need for soil topics and NSTA would be supportive of efforts to develop them. Dr. Gennaro also indicated that many of the earth science textbooks stress oceanography, but that for many of the interior states the study of soil science would be much more fitting. Textbooks often dictate what is to be taught, as was indicated by many teachers in our survey.

The recommendations that we have determined from our survey and interviews are centered around the development of curriculum and teacher training. Because many of the teachers in our survey have 10 yr or less before retirement, the opportunity to have more of our future teachers appreciate the use of soils concepts is there. This could be accomplished by having a soils course become one of the science electives in K-12 teacher curriculum.

RECOMMENDATIONS

1. Determine the procedures necessary to have soils concepts incorporated into K-12 curriculums:
   a. Invite teachers to ASA conferences and university departments to better acquaint them with the science of soils and to have dialogue with the scientists doing the science.
   b. Become acquainted with state and national curriculum development committees and with state department of educations that have a record of what is being taught in their schools.
   c. Become familiar with science teacher organizations like the NSTA, which will have affiliated state chapters. These organizations will have monthly or annual meetings for sharing activities and presentations by scientists and teachers.
   d. Visit with curriculum coordinators in local school districts who are often trying to find inservice ideas as well as new technology that can be taught in the classroom.
   e. Visit with individual teachers in local schools.

2. Develop and make available classroom activities, both hands-on and minds-on, that require small amounts of teacher preparation time. Teachers are looking for real life objects and demonstrations that are visible and exciting to their students. Publication of these activities can occur in science teacher professional journals like *The Science Teacher*, which is a NSTA journal. Their manuscript guidelines (Texley, 1992) are published monthly. They are interested in your first-hand experience that stress classroom applicability.
3. Provide introductory soils course work for both graduate and undergraduate credit for teachers. The graduate credit is especially important for teachers currently employed and will require offerings at night or during the summer. Introductory soils courses will provide the necessary information so that teachers can design their own soils activities for their classrooms.
4. Provide informal training for teachers. Activities could include field trips, inservice workshops, opportunities to work with scientists on research, and assisting their students with ideas for science fairs.

Increasing the flow of soils information from university soils programs to K-12 teachers will increase the amount of soils information currently being offered in our elementary and secondary schools. The long term effect of this should increase the number of college students wanting to learn more about soil science, regardless of the major they are studying in college.

REFERENCES

Barnes, R.F. 1987. Human resource needs, educational challenges, and professionalism in the agronomic sciences. J. Agron. Educ. 16:49-60.

Bugliarello, G. 1989. Physical and information sciences and engineering. Rep. of the Project 2061 Phase I. Am. Assoc. for the Adv. of Sci., Washington, DC.

Clark, M. 1989. Biological and health sciences. Rep. of the Project 2061 Phase I. Am. Assoc. for the Adv. of Sci., Washington, DC.

Litzenberg, K.K., D.A. Suter, and S.S. Whatley. 1991. Summary of fall enrollment in colleges of agriculture of NASULGC institutions. NACTA J. 35:4-11.

Texley, J. 1992. Write for the science teacher. The Science Teacher 59:95.

2 Undergraduate Core Curriculum in Soil Science

K. A. Barbarick
Colorado State University
Fort Collins, Colorado

ABSTRACT

Evaluation of course requirements, including core curricula, for all academic options is a continuous process at all institutions of higher learning. My objectives are to investigate the current status of course requirements in the soil science discipline at various institutions, to discuss the need for core curricula for various options in soil science, and to speculate on the background soil scientists will need in the future. Fifty-seven North American institutions that offer some type of 4-yr soil science option responded to an informational request by Soil Science Society of America (SSSA) Committee S571 (Training of Soil Scientists). The options available were placed in four categories; the distribution of the 78 curricular alternatives were: 10 in environmental soil science, 16 in soil resources, 37 in soil science, and 15 in *other* (generally encompasses aspects of soil management for crop production). Required coursework for these four groups of options were not significantly different for most types of courses. More credits of chemistry were required for environmental soil science than for soil resources options. In an apparent trade-off, the *other* option required more plant science, but less environmental or natural resources credits than the environmental soil science option. Based on the information provided to S571, a common core curriculum within the various options of soil science appears to exist already. I anticipate an increase in required credits involving interpersonal skills or communication and practical experience such as internships or research problems. Pressure to increase the number of required courses may result in 4.5- to 5-yr programs for soil science curricula.

Evaluation of curricula includes close scrutiny of required courses and examining the need for a core set of courses. Accrediting organizations usually mandate minimal course requirements or a core curriculum. More than likely, the development of core curricula for various disciplines will increase in the future. Need for a universal core curriculum in soil science is an intriguing concept.

Environmental concerns such as waste disposal and groundwater quality have undoubtedly led to more soil science courses that address environ-

---

Copyright © 1994 Soil Science Society of America, 677 S. Segoe Rd., Madison, WI 53711, USA. *Soil Science Education: Philosophy and Perspectives*. SSSA Special Publication no. 37.

mental concerns (Page & Letey, 1972; Barbarick, 1992) and subsequently to development of curricula with an environmental orientation (Letey & Page, 1972; Cooper, 1990; Daniels et al., 1992). Weis (1990) and Brough (1992) indicated that environmental-type programs are often described as *soft* sciences, since they tend to require less coursework in *hard* sciences such as chemistry, physics, and mathematics. Can the same statement be made for the burgeoning environmental soil science options? Discussion of the course requirements for various options in soil science is needed now.

My objectives are to: (i) examine the current status of course requirements at various institutions, (ii) address the need for core curricula for various options in the discipline of soil science, and (iii) ponder the background soil scientists will need in the future.

CURRENT COURSE REQUIREMENTS

Dr. Philippe Baveye, chairman of SSSA committee S571 requested information on curricular requirements from North American institutions that offer 4-yr programs in an option in soil science. Fifty-seven universities responded. Table 2-1 provides a listing of the schools that replied, the options offered, and a coding of the curricular choices. Four option codes were utilized: environmental options were Code 1; soil and water resources options (including irrigation) were Code 2; soil science options were Code 3; and other options such as agronomy and soil and crop management were Code 4.

Table 2-2 shows that the soil science curriculum is the most frequently listed selection; but, the number of environmental soil science options has undoubtedly increased during the last decade. This trend will continue since this curriculum probably has a broader appeal to prospective students and will probably lead to increased enrollment in departments offering this choice.

Average semester course requirements for the four options are presented in Table 2-3. Economics and agricultural economics requirements were placed in the social sciences category; foreign language requirements were placed in the humanities category. The *other* classification includes wellness and elective courses. Oneway analyses of variance were used to determine significant differences in average required credits among the four general options. Least significant difference at the 0.05 probability level was used to compare means.

Required coursework for the four options were not significantly different for the general topics of composition and speech (communications), humanities, social sciences, math, physics, biology, soil science, or *others*. The first seven categories of classes listed in Table 2-3 are generally considered *liberal arts* courses. Except for chemistry, all four options require essentially the same quantity of liberal arts credits. Out of an average semester credit requirement of 127 for graduation, liberal arts courses constitute 61 units or 48% of the total (Table 2-3).

Table 2-1. North American institutions responding to a request† concerning their curricular options in soil science.

Institution‡	Curricular option	Code§	Institution	Curricular option	Code§
Auburn University	Soil science	3	North Carolina State University	Soil science	3
California State University-Chico	Soils and irrigation	2	North Dakota State University	Soil conservation	2
Colorado State University	Soil resources and conservation	2	Ohio State University	Soil science	3
	Environmental soil science	1	Oklahoma State University	Agronomy	4
			Oregon State University	Soil science	3
Cornell University	Soil science	3	Pennsylvania State University	Soil science	3
Iowa State University	Agronomy science	4	Prairie View A&M University	Agronomy	4
Kansas State University	Soil and water conservation	2	Purdue University	Soil and crop science	4
			Rutgers University	Soil science	4
Louisiana State University	Soil science	3	State University New York	Environmental and forest biology	1
	Land and water resource management	2	Texas A&I University	Soils	3
McGill University	General soil science	3	Texas A&M University	Crop and soil science	4
	Soils and crops	4	Texas Tech University	Agronomy science	4
	Soil conservation	2		Industry and management	4
	Soil research	4	University of Alberta	Soil science	4
Michigan State University	Crops and soils	4	University of Arizona	Soil and water science	3
	Environmental soil science	1	University of Arkansas	Soil science	2
Mississippi State University	Soil science	3	University of British Columbia	Soil science	3
Montana State University	Soils and environmental science	1	University of California-Berkeley	Soil environment	1
	Land resources	2	University of California-Davis	Soil and water science	2
New Mexico State University	Soil science	3			

(continued on next page)

Table 2-1. Continued.

Institution[‡]	Curricular option	Code[§]	Institution	Curricular option	Code[§]
University of California-Riverside	Environmental soil science	1	University of Massachusetts	Soil science	3
University of Connecticut	Agronomy	4	University of Minnesota	Soil science	3
University of Delaware	Environmental soil science	1	University of Missouri	Agronomy science	4
University of Florida	Soil science	3	University of Nebraska	Soil science	3
	Soil technology	4	University of New Hampshire	Soil science	3
	Soils and land use	2	University of Saskatchewan	Soil science	3
University of Georgia	Environmental soil science	3	University of Tennessee	Soil science	3
			University of Wisconsin-Madison	Soil science	3
				Soil resource and land use analysis	2
University of Hawaii	Soils	1	University of Wyoming	Soil science	3
University of Idaho	Soil science	3	University of Utah	Environmental science	1
	Soil technology	4		Soils and irrigation	2
University of Illinois	Soil science	3		Soil science	3
University of Kentucky	Soil science	3			
University of Laval	Soil science	3			
University of Maine	Soil and water conservation	2	Virginia Polytechnic Institute and State University	Soil science	3
	Earth science	4	Washington State University	Soil science	3
	Soil science-sustainable agriculture	3		Soil management	2
University of Manitoba	Soil science	3		Soil resources and land use	2
University of Maryland	Soil science	3	West Virginia University	Soil science	3
	Conservation of soil, water, and environment	1			

[†] Information compiled from material requested by Dr. P. Baveye for Committee S571 (Training of Soil Scientists).
[‡] Number of institutions responding = 57.
[§] Options: Code 1 = environmental; Code 2 = soil resources; Code 3 = soil science; Code 4 = other (agronomy, soils and crops, crops and soils, soil research, soil technology, industry management, and earth science).

Table 2-2. Frequency of curricular options.

Curricular code	General option title	Frequency
1	Environmental Soil Science	10
2	Soil Resources	16
3	Soil Science	37
4	Other	15
Total		78

For the majority of institutions listed in Table 2-1, the liberal arts credits are mandated by university policy. By default, therefore, the average curricular option in soils does contain a core curriculum of liberal arts requisites.

Significant differences in credits in chemistry, plant or crop science, environmental or natural resources, and agriculture were detected (Tables 2-3 and 2-4). Although there was a significant difference in chemistry credits between environmental soil science (15) and soil resources (10), the other two options also tended to have fewer hours of chemistry requirements. Obviously the notion that environmental options contain less *hard* science than other options is mythical. The apparent exchange of credits among options is that the *other* option requires more plant or crop science and less environmental or natural resources credits than the environmental soil science option. Soil science also required less environmental or natural resource credits than environmental soil science. This difference signifies the desire to provide distinct training between the curricular options and reflects the one general area of coursework where departments are still able to change graduation requirements. The soil science option also required more general agriculture courses than the environmental soil science category. I believe that this difference represents a remanent of a more traditional agricultural approach to soil science curricula.

A core curriculum in the various options in the soil science discipline already exists. While differences in requirements in chemistry, plant or crop science, or environmental or natural resources exist between options, they represent a small fraction of overall course requirements (Table 2-4). Most of the significant shifts in course requirements compared with the traditional soil science curricula constitute <6% of the overall graduation credits. The exception is the 6.4% increase in environmental or natural resource courses required for the environmental soil science option. Because of university or college edicts, changes in options are limited to courses that may be described as restricted electives. In essence, soil science students, regardless of the option selected, apparently do and will have fairly similar coursework backgrounds.

The vast majority of undergraduate soil science programs encourage students to obtain practical experience before graduation. This is commonly accomplished through internships, work cooperatives, or independent-study research projects. Some institutions require this experience for graduation. Although this educational experience is desirable, adding credits of this type to existing requirements may create hardships for students. Meeting coursework demands of the university and department plus the need for practical

Table 2-3. Average semester units (standard deviation) required for various types of courses for the four general options.

General options	Programs†	Composition and speech	Humanity	Social science	Mathematics, statistics	Chemistry	Physics	Biology	Plant, crop	Environ. and natural resour.	Soils	Agriculture	Others	Total
	no.	Average number of credits												
Environmental soil science	8	8	6	9	7	15	6	9	5	13	17	0	29	125
Soil resources	13	10	7	10	7	10	4	10	8	11	18	2	28	125
Soil science	27	10	9	8	8	12	5	8	10	4	20	6	25	129
Other	10	11	8	11	9	12	4	9	16	5	15	3	25	129
LSD (0.05)‡		NS	NS	NS	NS	4	NS	NS	8	8	NS	6	NS	NS
Overall average	58	10	8	9	8	12	5	9	10	6	18	6	28	127
		(2)	(4)	(4)	(3)	(4)	(3)	(4)	(7)	(7)	(5)	(6)	(12)	(4)

† Exact credits for certain programs were either provided or could not be determined from the information provided.
‡ Least significant difference at the 0.05 probability level; NS = not significant.

Table 2-4. Changes in course requirements compared with the soil science option where significant differences between options were found.

General options	Chemistry	Plant, crop	Environment and natural resources	Agriculture
		—— % vs. soil science† ——		
Environmental soil science	+2.4	−4.0	+6.4	−4.8
Soil resources	−1.6	−1.6	+5.6	−3.2
Other	0	+4.7	+0.8	−2.3

† Percent change based on the total for each option and compared with the requirements in Soil Science, the most frequently listed curricular choice in Table 2-1.

experience could increase the total credits needed for graduation. These changes could result in B.S. programs that necessitate 4.5 to 5 yr to complete. I think that educators too often expect soil science students to accomplish too much within a 4-yr curriculum.

A student in soils must develop communication skills and a basic understanding of humanities, social sciences, and natural sciences along with a good blend of technical instruction. Selection of the right blend of coursework and practical experience within a 4-yr program is a very large challenge.

WHAT ABOUT THE FUTURE?

Increased demands by universities for more liberal arts in the curricula plus attempts to incorporate internship-type experiences will either increase total credits required for graduation or result in fewer restricted or free electives. Soil scientists, regardless of the curricular option completed, will work more as a part of interdisciplinary teams. Consequently, courses or other experiences that develop interpersonal skills should be encouraged.

Since environmental soil science options are increasing, the author believes that an increased emphasis on communication skills will be necessary. Regulatory agencies such as the U.S. Environmental Protection Agency will seek graduates who are articulate and who can handle presentations at public hearings and news conferences.

Institutional requirements will necessitate that shifts in coursework occur primarily within restricted or free electives. Environmental soil science options will continue to deemphasize traditional plant or crop science courses while stressing environmentally-oriented classes. This shift will undoubtedly lead to an increase in environmental-type soils courses.

Some courses with an environmental orientation are available (Barbarick, 1992). New courses in this area should use case studies as a primary emphasis. These capstone courses should build on the basic concepts learned in other courses and stress problem solving through real world examples. Role playing in these classes also may be important. Environmental soil scientists will probably interact a great deal with the general public; sometimes an adversarial situation may exist. Preparing students for various situations (e.g., public hearings) should be a key emphasis in these environmental courses.

An increase in the number of institutions that offer minors in various options in soil science will probably increase. Students from related disciplines in natural resources, such as range science, will increase their marketability as a scientist by developing secondary areas of expertise.

The future for all types of soil scientists is very bright. The need for individuals who understand environmental, agricultural, and resource management aspects of soils will certainly increase. Soil science educators will have the responsibility to provide opportunities to students so that they will eventually play a major role in society. The one constant for the future is that change will occur. Institutions of higher learning must allow modification of soil science curricula to address future needs appropriately.

ACKNOWLEDGMENT

The author thanks Dr. P. Baveye, chairman of S571, for providing the surveys on soil science curricula that individual institutions completed.

REFERENCES

Barbarick, K.A. 1992. An environmental issues in agronomy course. J. Natl. Resour. Life Sci. Educ. 21:61–63.

Brough, H. 1992. Environmental studies: Is it academic? World Watch 5(1):26–33.

Cooper, T.H. 1990. A natural resources and environmental studies curriculum. p. 3. *In* Agronomy abstracts. ASA, Madison, WI.

Daniels, W.L., J.R. McKenna, and J.C. Parker. 1992. Development of a B.S. degree program in environmental science. J. Natl. Resour. Life Sci. Educ. 21:70–74.

Letey, J., and A.L. Page. 1972. Problems in developing a curriculum in environmental sciences. J. Agron. Educ. 1:64–68.

Page, A.L., and J. Letey. 1972. Designing courses to meet the needs of environmental concerns. J. Agron. Educ. 1:68–72.

Weis, J.S. 1990. The status of undergraduate programs in environmental science. Environ. Sci. Technol. 24:8.

3 Trends in Soil Science Teaching Programs

J. Letey
University of California
Riverside, California

ABSTRACT

Traditionally undergraduate students majoring in soil science came from rural backgrounds with a career goal of employment in some phase of agriculture. Graduate students, particularly those with Ph.D. degrees, received employment in research or teaching at universities or with the Agriculture Research Service. Presently soil science teaching programs must attract students from urban areas and prepare them for successful careers in a broader spectrum than agriculture. Graduate programs should be adjusted to train soil scientists for positions in the private sector as well as the traditional positions. Teaching institutions must adapt undergraduate and graduate programs to accommodate societal shifts or face declining enrollments. Revision in curricula rather than merely relabeling will be required for continued success. Many institutions are adjusting to include an environmental focus to their soil science teaching programs.

The 20th century has been characterized by major societal transitions in the USA. Whereas most of the population lived in rural communities and small family owned farms at the beginning of the century, the vast majority of the population presently lives in urban areas. Major shifts also have occurred in rural areas, where there has been a transition from small family farms to larger corporative entities.

Soil scientists have served a significant role in this progression of events. Research into the fundamental principles of fertilization, irrigation, and land management practices has led to large increases in crop production per unit land area. Advances in agricultural production has freed much of society from the task of providing food so that they can pursue other productive ventures. Soil scientists have contributed to the educational programs for students interested in agriculture related careers, ranging from farming to research. Since initial emphasis was generally on agricultural production, soil scientists were often administratively combined with crop scientists into agronomy departments on university campuses.

Copyright © 1994 Soil Science Society of America, 677 S. Segoe Rd., Madison, WI 53711, USA. *Soil Science Education: Philosophy and Perspectives*. SSSA Special Publication no. 37.

The era of environmentalism evolved in the late 1960s and early 1970s. It was characterized by *Earth Days* and by the initiation of major federal legislation directed towards environmental quality. Society, being largely freed from the concern for food supply, focused increased attention on the environment. Agriculture was no longer viewed simply as an effective food and fiber supplier, but also as a contributor to land and water degradation.

Soil scientists, whose traditional research and teaching roles were associated with production agriculture, had a decision to make. Should they maintain the *status quo* or broaden their research and teaching programs to encompass new concerns? The question actually consisted of two separate but related issues. The first issue was whether environmental concerns as well as the production aspects of agriculture should be pursued. The second was that, since soil and water degradation was not restricted to agricultural activities, should soil scientists expand their role to include nonagricultural problems? One opinion was that, since many soil scientists were members of agricultural experiment stations or the U.S. Department of Agriculture, they were obliged to serve agriculture; to do research and teaching that was not agriculturally oriented was perceived to be a misuse of funds. A narrower attitude, held by some with long-term agricultural associations, was that it was not appropriate to engage in environmentally oriented activities even if they were related to agriculture, since this might uncover information that would be detrimental to the agricultural industry. It was not uncommon for the first scientists to report quantitative data on water degradation by agriculturally-used chemicals to be criticized by some of their scientific colleagues. Nevertheless, within the last two or three decades there has been a large change in the type of teaching and research performed by soil scientists. It now encompasses a full range of issues, including agricultural production and environmental quality in both agriculturally and nonagriculturally related settings.

Recent expansion in soil science research has been largely driven by the availability of extramural funding sources for nontraditional agricultural research. Furthermore, soil scientists have recognized that they were doing basic research related to the fate and transport of chemicals in soil and water systems, and that these basic principles had broad application to both agricultural- and nonagriculturally oriented settings.

Conducting research into nontraditional agricultural problems brought soil scientists into interaction with, and sometimes into competition with, other disciplines including engineering, geology, and geography. Soil scientists have both advantages and disadvantages when competing with other disciplines in nonagricultural activities. The advantage is that soil scientists have training, or colleagues with training, in all major aspects important to the fate and transport of chemicals in soil–water systems. Chemical, physical, and biological aspects can all be addressed via a well-integrated program. The major disadvantage has been the failure of some individuals, particularly in the private sector, to recognize the expertise and capabilities that soil scientists have at their disposal. This disadvantage is becoming less important as the contributions of soil scientists become more widely recognized.

TRENDS IN SOIL SCIENCE TEACHING PROGRAMS

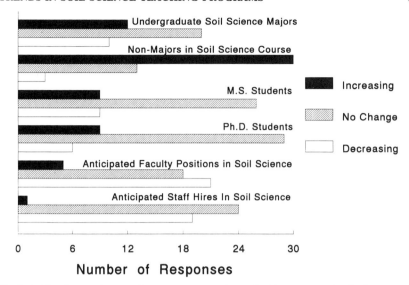

Fig. 3-1. Numbers and types of responses to questions asked in a survey of soil science teaching programs.

Transitions in the research agenda for many soil scientists can be documented by publications and reports at scientific meetings. Transitions, if any, in teaching programs are less evident. In order to more accurately quantify what is happening with teaching programs, a questionnaire was sent to institutions in the USA and Canada that teach soil science. A total of 46 out of 60 questionnaires were completed and returned. One part of the questionnaire asked the respondent to classify recent numerical trends for various items as either increasing, decreasing or remaining about the same. Items listed on the questionnaire, and a summary of responses, are presented in Fig. 3-1. Institutions were also asked to: (i) briefly describe changes in undergraduate courses or curricula that have been made during the past few years, (ii) briefly describe changes in graduate courses or curriculum that have been made during the same period, (iii) describe present plans for future modification of soil science teaching programs, and (iv) provide any additional thoughts on the subject of soil science teaching program that they would like to share.

UNDERGRADUATE EDUCATION

Almost without exception, each respondent identified a shift in undergraduate soil science to increased environmental or resource emphasis. In some cases, *new* undergraduate soil science programs with emphasis on the environment have been developed. Some actually carry the tile *environmental soil science*. In other cases, soil science courses are part of a campus- or college-wide environmental science program, or have been specifically tailored to attract nonmajors. Details of trends and transitions in the soil science program at California Polytechnic State University, San Luis Obispo, are presented in an accompanying article by Taskey (1994) as an example of what has been done at one institution.

Based on the data presented in Fig. 3-1, most institutions have been successful in attracting more nonmajors into soil science courses by introducing greater environmental orientation. The number of respondents reporting increased numbers of undergraduate soil science majors is approximately equal to those indicating decreased numbers of majors. Part of this trend may be attributed to the fact that some departments have dropped the soil science major, but have concurrently become an integral part of a broader environmentally oriented program. One respondent stated that the traditional undergraduate degree is a thing of the past. Another respondent stated that institutions have abandoned the philosophy that all undergraduate majors need to have exactly the same type of soil science background.

Although only one department commented on the following issue, it is probably an issue that most departments are facing nationwide. With declining resources it is imperative that each department be competitive in attracting students. Administrators are evaluating the college-wide or university-wide allocation of teaching resources based on student numbers. Whether the budget is a primary motivating factor, it is obvious that significant restructuring at the undergraduate level has been done at most, if not all, institutions and that such efforts have been generally successful in attracting more students.

GRADUATE TRAINING

Whereas almost every respondent identified major shifts in the undergraduate teaching program, a common response at the graduate level was that relatively few changes have been made. The following response is typical of most respondents: "Changes in our graduate curriculum in soil science have occurred mainly through the emphasis of our research program which is dealing more with environmental contamination, reactions of wastes and waste products in soils, etc. One new course has been added about 2.5 yr ago." Apparently most departments have seen little need, or at least have not invested the time required, to make major modifications in their graduate teaching program, other than as reflected by research orientation.

Approximately equal numbers of respondents identified increased graduate student numbers as reported decreased graduate student numbers, with the majority indicating that there has been relatively little change.

JOB OPPORTUNITIES

One of the respondents stated, "Students today come into the program with the following question: 'What kind of job will I get when I get out?' They want the curriculum to be specific to their needs. Perhaps this could be characterized as trade school mentality, but job opportunities are more and more important to them." This statement was made relative to an undergraduate program. This question from undergraduate students is undoubtedly asked at most institutions, and has probably contributed to the above-mentioned modification of undergraduate curricula.

At the graduate level, the data reported in Fig. 3-1 are disturbing. With regard to anticipated faculty positions in soil science, almost one-half of the institutions indicated a decreasing trend. Only a few indicated an increasing trend, and most of these institutions were those with a relatively small number of soil science faculty. One institution indicated that they have been authorized faculty replacement for only one of every two open positions, and that this could soon go to one of three. Note also, in Fig. 3-1, that the prospects for anticipated staff hires in soil science is rather bleak as well. If these projections are accurate, traditional job opportunities for Ph.D. and M.S. soil science graduates at universities will be greatly diminished. One conclusion is that Ph.D. students in particular must be trained to be competitive in a broader job market than university teaching and research. In particular, the private-sector job opportunities must be tapped.

Two issues must be addressed when expanding the job opportunities in the private sector for M.S. and Ph.D. students. The first question is whether traditional graduate training is the most appropriate training for indivdiuals going into the private sector. Secondly, the private sector needs to be made aware of the capability of M.S. and Ph.D. recipients in soil science to meet their needs.

Many private sector positions are likely to come from nonagricultural entities. Traditionally, soil science has been combined with crop science to optimize the capability for addressing agricultural production problems. With the broadening scope of soil science research and teaching, however, one must question whether continued alliance with crop science is the optimal arrangement. On the questionnaire, one respondent indicated that soil science has now been moved out of agronomy to the School of Natural Resources. Another respondent indicated that there had been a suggestion to combine hydrology from agricultural engineering and geology with soil science to form a Land Resources Department. Others at the same institution, however, suggest that agronomy should maintain and strengthen, rather than break, the ties between soil and plant sciences.

Soil science appears to have positively and effectively responded to changing times with regard to research and undergraduate teaching. Thus far, however, few if any modifications appear to have been made in graduate teaching programs other than research-topic emphasis. Each department offering an M.S. or Ph.D. degree in soil science must address the following questions: Can the *status quo* be viably sustained into the future? If the *status quo* does not appear to be feasible and the projected trend in hiring Ph.D.s into traditional teaching and research continues, then what changes must be made? An obvious answer is to train students for a broader job market. As each job market is identified, appropriate and competitive training must be identified. This might require restructuring of subject matter within a revised set of courses, consideration of using internships with the private sector as a significant training component, introducing quite different subject matter, and possibly considering a restructuring of the administrative structure to enhance the training of soil scientists for nonagricultural positions.

One lession that can be learned from ecology is that species that are unable to adapt to changes in their environment become extinct. Soil scientists have demonstrated that, in research and undergraduate teaching, they have remained flexible and in general prospered. Based on this track record, one can hope that graduate program adjustments will follow as well, and that the soil science profession thus will flourish. This positive outlook, however, is premised on the assumption that soil scientists will invest the time and effort to respond to emerging needs at the graduate level.

REFERENCE

Taskey, Ron. 1994. Revision and rescue of an undergraduate soil science program. p. 21–27. *In* Soil science education: Philosophy and perspectives. SSSA Spec. Publ. 37. SSSA, Madison, WI (this publication).

4 Revision and Rescue of an Undergraduate Soil Science Program

Ronald D. Taskey
California Polytechnic State University
San Luis Obispo, California

ABSTRACT

The undergraduate soil science program at California Polytechnic State University, which has been among the largest in the nation for more than two decades, suffered a severe decline in enrollment in the middle to late 1980s. Departmental autonomy became seriously threatened, and, more importantly, the program was identified for possible elimination. The faculty responded by establishing three new concentrations under the soil science degree program: land resources, environmental management, and environmental science and technology. Overall, the new program, which was created solely from existing resources, is more rigorous than the traditional curriculum. Potential new students were invited to apply. As a result of these concerted efforts, soil science enrollment nearly tripled within 2 yr.

One undergraduate program in soil science that recently was redefined and restructured to meet emerging societal needs is that at California Polytechnic State University, San Luis Obispo. Cal Poly's undergraduate soil science program has been among the nation's largest for at least two decades. Its traditional role has been to educate students for positions in soil conservation, soil survey, soil and plant analysis for agriculture, the fertilizer and agricultural chemicals industries, farm advisement, and land reclamation; and for graduate studies.

Although enrollment had been strong since the program's inception, it increased dramatically in the early 1970s (Fig. 4-1), reflecting the country's new-found interest in the state of the earth. Following the first Earth Day, students were attracted to ecology and natural resources programs in large numbers (U.S. Department of Education, 1992). These people knew that they wanted to do something for the earth, but many of them lacked clearly defined goals. Moreover, many of the newly developed or reorganized programs in which they enrolled lacked the wherewithal to progress much beyond

Copyright © 1994 Soil Science Society of America, 677 S. Segoe Rd., Madison, WI 53711, USA. *Soil Science Education: Philosophy and Perspectives.* SSSA Special Publication no. 37.

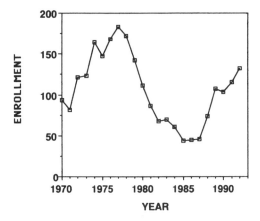

Fig. 4-1. Cal Poly undergraduate soil science enrollment, 1970 to 1992.

problem recognition and analysis: college environmental programs offered few real social, economic, or technical solutions to environmental problems. As a result, programs flourished, but opportunities for environmental generalists were limited, and the job market soon became saturated.

Students responded by shifting to more specialized environmentally oriented programs, such as forestry, hydrology, low input farming, and soil science. Accordingly, Cal Poly's Soil Science Department welcomed new students eager to transfer from community colleges and 4-yr programs that they found too general or ill-defined. By the late 1970s, the program supported ≈175 undergraduate students, and the introductory soils course bulged with nearly 1200 students per year. Meanwhile, the number of faculty doubled from 6 to 12 full-time teachers.

The typical soils student changed also. Until this time, the soil science student body had been strongly dominated by white males from rural backgrounds, with interests in farming and related support professions. The new students brought more diversified interests in forest and urban land use, organic and low input farming, and soil quality degradation, and many discovered satisfaction in traditional soil science. Moreover, the proportion of female students increased from <10 to >40%. Women came to power quickly, both in the classroom and in departmental activities.

Amid all the expansion and changes, the curriculum remained virtually static. Faculty and administration saw no reason for revision, only for a few minor adjustments to accommodate specific interests of new teachers. After all, they reasoned, the prevailing curriculum had served the school and profession well, and enrollment was greater than in any other comparable program in the country, and probably the world! So, although a few faculty may have sensed an impending barometric change, the decision was to stay the course.

But as rapidly as the tides of students and new opportunities rose, they likewise ebbed (Society of American Foresters, 1991; U.S. Dep. of Education, 1992) when governmental and popular philosophy shifted to increased laissez faire and decreased support for agriculture, natural resources, and the environment. Governmental moves to deregulate and disencumber led to large profits in the business sector and volunteerism in the public sector. The new-found wealth of business and industry trickled in directions other than those in which agriculture and natural resource students were headed.

As a result, Cal Poly's student boom in soil science lasted slightly less than two student generations. By 1985, enrollment had dropped precipitously, to only 44 majors and ≈ 600 students per year in the introductory soils course, considerably lower than the enrollment even before the 1970's boom. Although these fluctuations were extreme, this mid-1980's enrollment decline coincided with national trends of students away from agricultural and natural resources fields (Society of American Foresters, 1991; Food and Agricultural Education Information System, 1992).

The enrollment remained between 44 and 46 through 1987, during which time only a few applications were received. The program and the profession it served were in trouble. The Soil Science Department, along with other similar departments in the country, were threatened with loss of departmental status, and, more importantly, with loss of the program. The advice offered by university administrators was to recruit new students, in effect to take a show on the road to high schools, community colleges, and even county fairs and trade shows. Some suggested that expectations placed on students were too high, and that these might be adjusted to make the program more attractive. (Of course this somehow would be done while maintaining the widely recognized high standards of the university system.)

The recruitment suggestion was rejected on the grounds that it simply would be using feigned enthusiasm to create a false demand. The entire profession, not simply university enrollments, was in a slump; it would be improper to lure people into a major that held little promise for career opportunities upon graduation. Nonetheless, the faculty believed in the need for a strong undergraduate soil science program, especially in a large farm and natural resources state such as California. They also believed that many new, nontraditional, opportunities could be available for graduates if the program was redesigned and the rest of the world discovered soil science.

GOALS, CRITERIA, AND METHODS

After much deliberation and consultation with alumni and representatives of industry and governmental agencies, the faculty gradually developed the goals, criteria, and methods needed to expand the academic offerings. The goal that finally emerged was to develop a tripartite curriculum that would (i) take advantage of traditional opportunities and offer students diversity and flexibility; (ii) prepare soil scientists to become effective in resource and environmental planning, policy, and administration; and (iii) provide a strong scientific foundation upon which graduates could compete for rigorous graduate programs and for positions in land and water pollution abatement.

In addition, they set criteria and recognized constraints as follows:

1. Maintain traditional program goals and commitments to clients in agriculture and natural resources; to do otherwise would debase the recognized importance of soil science in food and fiber production.
2. Provide an education that will be highly valued by society over the long term. A curriculum strong in fundamentals will allow graduates the greatest flexibility in the future. Students will develop a firm foundation in general education, basic sciences, mathematics, and soil science.
3. Graduates must have high probability of finding immediate and long-term professional opportunities, even if they are not employed directly in soil science.
4. Minimize the number of courses students are required to take from the Soil Science Department to ensure that they are exposed to a wide variety of disciplines and faculty. This criterion reflects a philosophy quite different from that of most other departments, which tend to internalize their students, especially when enrollments are low.
5. Any changes must be made without an increase in resources, faculty, or staff; only those resources available at Cal Poly could be used.
6. Any restructuring must be supported by the rest of the university (most notably, of course, the curriculum committees at college and university levels, and academic senate, each of which have curricular authority), alumni, industry, governmental regulatory and land management agencies, and other universities, especially those having significant graduate programs.
7. Department faculty must share equally in administering any new concentrations; no faculty member may be identified as *in charge of* or advisor to one concentration in preference to any other concentration. Cooperation must be emphasized; territorialism must be discouraged.
8. Well qualified potential students must be available to fill the program.

The goal was accomplished by following several steps, some planned, others fortuitous:

1. Identify new job markets and professional opportunities for graduates.

    These were becoming available as Cal Poly soil science graduates increasingly found employment with private consulting firms and governmental agencies to work on various aspects of land degradation by hazardous wastes. (One firm hired nine graduates during the 3 yr the new program was being developed.)
2. Secure the support and counsel needed to develop the curriculum, have the program approved, and make it work.

    Other departments at the university, including some in the College of Agriculture, might be concerned that efforts in soil science would take away students from their programs, or that the Soil Science Department was becoming less supportive of their programs by changing requirements in a way unfavorable to them. A strong attempt was made to anticipate these concerns and alleviate them as efforts progressed. Fortunately, enough encouragement and good will were received from outside the university that other departments became convinced that the changes would be for the good of all.

Letters of support were amassed from alumni; graduate schools; industry (mostly environmental engineering and soil analysis firms); representatives of local, state, and federal governmental agencies, including the County Environmental Coordinator's office, State Departments of Health Services and Food and Agriculture, U.S. Forest Service, Soil Conservation Service, and Environmental Protection Agency. The proposed concentration in environmental science and technology was strongly endorsed in 1988 by the Soil Science Department *Five Year External Review* (Ludwick et al., 1988, unpublished data). Moreover, representatives of each of the groups mentioned reviewed draft proposals and made recommendations.

In 1988, Cal Poly's student chapter of the Soil and Water Conservation Society hosted the annual California conference, with the theme *Hazardous Waste in California's Soil and Water*. The meeting brought in influential speakers and attendees who provided valuable advice, supported the efforts, and offered jobs to students.

3. Identify courses and resources, including faculty, facilities, and equipment, at Cal Poly that were not being utilized by soil science, and which could help strengthen and expand the program into three concentrations.

The land resources concentration was developed from the traditional curriculum by rearranging certain support courses and providing students a sizable block of restricted electives from which to choose. Students thereby are allowed greater freedom in selecting their course of study, and in pursuing any of several minors offered by other departments.

The environmental management concentration was added through an agreement with the Natural Resources Management (NRM) Department to dual-list the program, which that department had offered for several years. The agreement opened another academic opportunity for soil science students and increased the enrollment in NRM classes.

The environmental science and technology concentration was created with the cooperation of the Departments of Chemistry, Physics, Mathematics, Statistics, Agricultural Engineering, and Environmental Engineering. This is the most scientifically demanding of the three concentrations (Table 4-1).

4. Locate and recruit potential new students.

As one of the most popular campuses in the California State University system, Cal Poly must deny admission to many well qualified applicants each year because the degree programs to which they

Table 4-1. Quarter units in major and support categories in each of the three concentrations.

Subject areas	Land resources	Environmental management	Environmental science & technology
Soil science	43	43	43
Biology, crop science	16	12	12
Geology	7	7	7
Chemistry, physics	24	24	39
Math, statistics	14	13	18
Irrigation-hydrology	4	4	3
Computers	3	3	3
Environmental analysis	--	10	--
Law	--	6	--
Planning, administration	--	9	--
Internship	--	3	--
Environmental engineering	--	--	5
Restricted electives†	23	--	4

† Students may choose restricted electives from a list of ≈100 courses, which includes those required for a minor in another discipline. Soil science courses are not included, even though several additional soil science courses are taught.

apply are full. These programs are those that are well known by guidance counselors and the public at large.

With the cooperation of the university admissions director, the dean, and appropriate department head in the College of Science and Mathematics, the Soil Science Department invited applications from those biological science applicants who were unaccommodated but well qualified, and who, on the basis of their application statements, might have an interest in soil science if they knew about it. As a result, 32 new students were admitted that year, increasing the enrollment from 44 to 76.

The faculty anticipated that most of these students would transfer out of soil science at their first opportunity, but this did not happen; nearly all stayed in the program. Most of these new students were entering freshmen from metropolitan areas—a very different student population from that of a few years earlier.

CURRENT SITUATION

The goal was to increase the soil science enrollment to 90 students 5 yr after implementing the revised program. By the end of the first 2-yr catalog cycle (1990–1992) the enrollment had reached 120 students, all of whom are expected to have professional opportunities upon graduation. The faculty are confident that the enrollment could have been raised to 200 students or more if the great budget crisis of 1992 had not struck, forcing a great loss of resources and a cap on enrollment. Nonetheless, the Department began the Fall 1993 term with 160 undergraduate students, including several who had transferred from chemistry, biological sciences, natural resources, and, for the first time, environmental engineering. These students were attracted, not by a facile curriculum and promise of easy grades, but by a demanding program, one designed to serve them well over the long term, and by a friendly, open faculty committed to helping them excel in a competitive world.

WHAT NEXT?

Once new directions have been charted and new programs have been implemented, the faculty must make them work. Contacts with industry, government, universities, alumni, and kindred professions must be strengthened and continually reinforced. Promising new students must be recruited and made to feel welcome. Current students must be nurtured. Standards must be kept high, firm, and reasonable. A dynamic unity must be maintained among faculty. Professional certification and registration programs must be supported so that future soil scientists will be fairly recognized to do work for which they are qualified.

Finally, soil scientists must strive to enhance the integrity and flexibility of their profession. University faculty must resist pressures to dilute the curriculum, or to surrender professional recognition in the name of reorganization. All soil scientists must work to create new opportunities and to retain ownership of their expertise. If the soil science profession is to remain viable and dynamic into the coming millennium, opportunities and expertise must not be forfeited or relinquished to other professions, either willingly or by default.

REFERENCES

Food and Agricultural Education Information System. 1992. Fall 1991 enrollment in agriculture and natural resources. Combined report for the Am. Assoc. of State Colleges of Agric. and Renewable Resour. (AASCARR) and Natl. Assoc. of State Univ. and Land Grant Colleges (NASULGC). FAEIS, Texas A&M University, College Station, TX.

Society of American Foresters. 1991. Forestry enrollments rise. J. Forestry 89(11):42.

U.S. Department of Education. 1992. Digest of education statistics. Office of Educational Research and Improvement. NCES 92-097. Natl. Center for Education Statistics, Washington, DC.

5 Understanding Cognitive Styles: How to Teach to the Whole Soil Science Classroom

Diana B. Friedman
University of California
Davis, California

Rodney J. Parrott
Cornell University
Ithaca, New York

ABSTRACT

Many analytical models exist in the field of education to explain differences in the way that individual students learn. One area of research that has been studied extensively is cognitive styles, specifically field independence–dependence (FI-FD). The FI-FD model encompasses a continuum of learning approaches that has specific applications to making teaching more inclusive in the science classroom. Currently, many science courses are taught in a manner that primarily favors field independent students hence possibly discouraging field dependent students from participating in or succeeding at science. The FI-FD model has specific applications to soil science; by nature of its cross-disciplinary and practical applications to real life problems, soil science often attracts students from a wide range of backgrounds, many of whom may be more field dependent than students from traditionally *hard* science backgrounds. Increasing diversity in the classroom in the 1990s also indicates that there may be more of a need for FD instruction. We will explain how the FI-FD model can be successfully applied to the soil science classroom by: defining cognitive styles, explaining how to identify what type of learners you may have in your classroom, outlining how to determine what type of an instructor you are, and evaluating a soil science syllabus that will show how to teach to a range of students.

Researchers in the field of education have shown that students receive and process information in numerous ways and have developed many different models to explain these various approaches to learning. One widely employed model, cognitive styles, specifically the field independent–dependent construct, is particularly useful for understanding learning styles and hence improving teaching for many reasons; students and teachers can be easily assessed for their preferred style, it is well accepted [almost 4000 studies have been conducted on the subject (Moran, 1985)], it can be applied cross culturally, it parallels many other educational theories, and finally, once understood, it is easy to incorporate into classroom instruction.

For the most part, FI teaching strategies, which include analytical, impersonal, and abstract approaches to learning, have dominated in university science classrooms across the USA. The soil science classroom is no exception, and in the past, this approach has proven relatively successful with a population of students who have traditionally been white, male, and from rural and farm backgrounds, with a strong training in agronomic sciences.

Today, however, the soil science classroom is quite diverse and the FI model may no longer be appropriate. Students taking soil science may be from cities, they are female, they often come from foreign countries, and they are of various ethnic backgrounds. Many of the current students in soil science also come from a wide range of academic backgrounds such as environmental science, natural resources, forestry, science and technology, biology, or ecology, where previous instruction may have been less FI. As all of these groups become more prevalent in soil science classes, the needs of this changing student population must be addressed; instructors who are concerned with improving the quality of undergraduate education can begin to meet this challenge by employing more field dependent strategies.

We will explain that FI-FD is a useful model for teachers to understand and incorporate into classroom instruction. Although simple, the FI-FD construct enables instructors to utilize strategies that allow them to teach to a broader range of students, while still maintaining a customarily high level of educational instruction. By including an FI-FD approach to their teaching, instructors will hopefully facilitate the learning of a greater range of students.

INTRODUCTION TO COGNITIVE STYLES AND FIELD INDEPENDENCE-FIELD DEPENDENCE

Research on cognitive styles began in the 1940s and grew out of physiology research on perception. The original impetus came from the observation during World War II that some fighter pilots, when losing sight of the ground, would fly upside down or sideways (Ramirez & Castaneda, 1974). In daily experience, two standards normally work together directing a person to the same upright. First, a person generally knows which way is up based on cues from the visual environment, such as vertical lampposts or vertical door jams in the room. Second, cues are also received from internal sensations as one's body adjusts to gravitational pull in an atttempt to stay upright and in balance.

Curious about the flying pattern of these pilots, and the dynamics of each of these standards in influencing orientation to the upright, Herman A. Witkin, a psychologist, developed laboratory tests to measure how people locate the upright in space. (Witkin, 1978). Eventually, three tests were developed: rod and frame, body alignment, and rotating room (Witkin, 1949, 1952; Witkin & Asch, 1948).

In the rod and frame test, a subject sits on an upright chair in a totally dark room. The subject sees only a luminous rod inside a luminous square frame, both of which pivot around the same center. The rod and frame are tilted at various angles relative to each other. The subject's task is to move the rod until he or she experiences it to be in the true upright, while the frame remains in its initial tilted position. In the body alignment and rotating room tests, a subject is placed on a chair in a room and either the chair or the room is tilted at various angles. The subject is then requested to align his or her body with the true upright.

Experiments over the course of many years yielded an extreme range of responses from subject to subject. The results seemed to indicate that, when presented with the perceptual paradox of visual cues from the environment that contradicted the bodily cues from within, many people chose one set of cues and denied the other. Witkin concluded that people seem to have a preferred style of perceiving that is utterly compelling and difficult to overcome (Witkin, 1978).

Furthermore, research showed that particular subjects consistently preferred the same style of perception in all three tests. A subject (*Alpha*) who experienced the rod as vertical only when aligned with the frame was also likely to tilt his or her body far off true upright in order to align with the tilted room and, predictably, have no difficulty in adjusting his or her body to the true upright when the chair was tilted. In contrast, another subject (*Omega*) who adjusted the rod to the true upright regardless of the position of the frame also was likely to align his or her body with the true upright regardless of the tilt of the room in which he or she was sitting, and again predictably, he or she would experience his or her body as vertical on the tilted chair even though its whirling forced his or her body far off true upright (Witkin et al., 1977, p. 6).

The key for Witkin that provided a synthetic explanation of an individual subject's consistency of perceptual style was the essential and protean notion of embeddedness. In all three of Witkin's laboratory tests, the subject was forced to employ a perceptual strategy in regard to disembedding. For example, in the rod and frame test, the rod is embedded in the field created by the tilted frame and must be isolated from that field in order to adjust the rod to true upright. In the body alignment test the subject him or herself, i.e., the body, is embedded in the visual field of a tilted room and the internal cues of bodily sensations must be disembedded from the cues of the external environment in order to move the body to true upright. In the rotating room test, again, the body is embedded in the visual field of the room, however, in this case, the body must remain embedded in the visual field in order to find the true upright.

A later test also developed by Witkin most clearly illustrates the notion of perceptual embeddedness: the embedded figures test. This test is a paper and pencil assessment that presents subjects with a series of complicated geometric figures in which a simple figure is hidden. The task is to find the simple figure and abstract it from its complicated field. *Alphas* have difficulty disembedding the simple figure, whereas *Omega* easily disembed the simple figure. Witkin posited that *Alphas* were exhibiting their consistent style in which perceptual experience was dominated by the organization of the field as a whole. The part, i.e., the embedded simple figure, was experienced as fused to the field as a whole, and was thus not perceived as isolatable. *Omegas* were exhibiting their consistent style in which perceptual experience was dominated by the distinctness of the parts, and thus the simple figure was isolatable from the whole. Statistical agreement across the four tests discussed is somewhat variable, yet always significant (Witkin et al., 1971; Ramirez & Castaneda, 1974).

For people akin to *Alpha*, who tended to prefer visual cues from the environment over bodily cues and perceived parts embedded in the whole, Witkin eventually coined the descriptive category *field dependent cognitive style*. People akin to *Omegas*, who tend to prefer internal sensations over visual cues from the environment, and who perceive parts distinct from the whole, Witkin described as exhibiting a *field independent cognitive style*.

Educational relevance to Witkin's studies grew out of his own realization that an adequate understanding of a person's perceptual style "could not be achieved without putting the person's characteristic way of processing information into the formula..." (Witkin, 1978). In other words, a person's perceptual style was based on how he or she approached learning and information, i.e., his or her cognitive style. Witkin eventually adjusted his research to this link between perception and cognition and conducted numerous studies that showed that the way in which people process information from their immediate environment and their bodies was also revealing the way in which people processed nonimmediate or symbolic information, i.e., cognitive representations (Witkin et al., 1971).

So eventually, although Witkin's research found its creative seed in the laboratory study of perception, he did not coin the phrase as *field dependent–independent perceptual* style, but rather as *cognitive style*. After years of research, Witkin concluded that cognitive styles are the characteristic, self-consistent modes of functioning that individuals show in their perceptual and intellectual activities (Witkin et al., 1971).

During the last 40 yr, volumes of research have been published linking field independent and dependent perceptual styles with cognitive styles. Cognitive styles are critical to the field of education in that they determine how individuals interpret and approach problems, and how they process information (Witkin et al., 1977). For example, *Alphas*, or field dependent people, learn material better that has a contextual basis, take a holistic approach to problem solving, and prefer structured material. *Omegas*, or field independent learners, prefer to think through new material alone, and to structure their own material. Field dependent *Alphas* prefer group work, aided by a narrative from a personable teacher. Field independent *Omegas* prefer abstract ideas, delivered in a lecture by a teacher at a professional distance (Witkin et al., 1977; Anderson, 1988; Davis, 1991). Appendix A lists more characteristics of FD and FI learners.

APPLYING COGNITIVE STYLES TO THE CLASSROOM IN THE 1990S

In *New Directions for Teaching and Learning*, Anderson and Adams (1992) indicate that more attention than ever is being focused on how to meet the challenge of increasing diversity in the classroom. "One of the most significant challenges that university instructors face is to be tolerant and perceptive enough to recognize learning differences among their students. Many instructors do not realize that students vary in the way they process and understand information. The notion that all students' cognitive skills are identical at the collegiate level smacks of arrogance and elitism by sanctioning one group's style of learning while discrediting the styles of others." (Anderson & Adams, 1992).

Research has shown that cognitive styles are not biologically determined but rather socially constructed. Since earliest information processing skills are taught to individuals by primary caregivers, one's cognitive styles are culturally conditioned (Anderson, 1988). Societal expectations and perceptions may also have an impact on how one is socialized to learn, especially once children enter the school system. Thus some researchers believe that FI-FD styles often break down among ethnic and gender lines; minority learners and women process information more along FD styles, while white males tend to be more FI (Ramirez, 1982).

Much research has been conducted to study whether ethnic minority learners are more FD than anglo learners. For example, some work has shown that African-Americans gain knowledge more effectively through tactile senses and verbal descriptions, and are socialized to concentrate on people rather than nonpeople types of information (Shade 1984). They also have the ability to succeed better working in groups (Shade, 1984). Ramirez and Castenada (1974) reported that Chicano students also indicate a preference for FD cognitive strategies such as relying on holistic skills, unlike middle class Anglo children, who have a preference for FI strategies.

If this is the case, then cognitive styles used by ethnic minorities are somewhat incompatible with current pedagogical practices, especially in math and science in the U.S. school systems, since so much of the teaching is FI. Ethnic minority learners specifically have a hard time trying to fit into the science domain because the learning in science requires analytical skills, abstract and impersonal orientation, and independent work. Consequently, the current methodology of teaching science attracts few minority learners.

This reasoning could well explain the high attrition rate of minority students in math and engineering at large universities, as well as account for the high retention rate of programs that emphasize FD instruction. Successful programs include the Academic Excellence Workshop at Colorado University, which encourages group learning, personal interaction and role modeling to help minority students succeed in math and sciences (Scott, 1991), and the premed-science program at Xavier University, a predominantly African-American school, which has emphasized cooperative learning strategies (Anderson, 1988).

Much research during the past 40 yr, has shown that women in the western world, upon reaching adulthood, tend to exhibit FD styles of learning, whereas adult men exhibit FI styles of learning (Demick, 1991). This research is further complicated by the fact that women, while having to contend with the above noted learning challenges presented by the FI science classroom, must also face a whole range of culturally created impediments associated with gendered role expectations. In this culture, an incongruency exists between the role of being a true scientist and being a true woman. Scientists are objective, logical, and impersonal, whereas women are contextual, intuitive, and personal. Thus, by the negative force of these role expectations, women at an early age are discouraged from the pursuit of science education beyond the introductory level (Fennema & Ayer, 1984). And, if certain enduring women do manage to enter collegiate science majors, they will, by the negative force of FI science pedology, be constrained in the expression of the FD cognitive style.

ROLE OF COGNITIVE STYLES IN SCIENCE EDUCATION

A wide range of studies show that FI students overwhelmingly do better in hard science classes such as chemistry, mathematics, physics, and biology. (Fields, 1985; Niaz, 1987, Davis, 1991). This issue, however needs to be framed in terms of the old *chicken and the egg* question. Do FI students do better because they are more intelligent and have a higher aptitude for science? Or rather, do they succeed more readily because in most educational settings the science curriculum favors FI learners?

This question itself is difficult to answer, so researchers have studied the relationship between cognitive styles of instructors and students to see if matching would improve FD student's performances in the classroom. For the most part, results indicate that pairing of FI and FD teachers with like students does not necessarily enhance learning. McDonald (1984) showed that FI-FD matching of students and teachers would only benefit a small number of college students, while Garlingher and Frank (1986) in a review of experiments on the subject, concluded that there was only a slightly higher level of achievement when students and teachers were matched. Riding and Boardman (1983) and Mahlios (1982) showed no improvement from matching.

Research has shown, however, that when teachers become aware of their cognitive styles and adjust their teaching methods—not their actual cognitive styles—to teach to both FI and FD students, both types of students show better performance. For example, Frank (1984) found that FD students who were given a structural outline of lecture notes (advanced organizers) from a beginning educational psychology class improved on a multiple choice test, as opposed to those who only had their own notes to study from. Crow and Piper (1985) showed similar improvements in a college geology class where the treatment group of FD students were (i) shown slides of geological features after being given verbal definitions, (ii) saw outlines drawn over the slides and projected onto the chalkboard, and then (iii) viewed the slides again. The control group of FD students received standard instruction of verbal definitions and slides and showed nonsignificant improvement. These studies suggest that it is not so much the actual personal cognitive style of the teacher that influences the success of the instruction, but rather, how that teacher designs his or her own instruction to meet the needs of both types of learners, that has a positive influence on the learning of the students.

HOW TO TEACH TO THE WHOLE SOIL SCIENCE CLASSROOM

Teacher Activities

In order to teach to the whole classroom, an instructor must be willing to assume that he or she is faced with a wide range of FI and FD learners in the classroom. Although research has shown gender and ethnic differences, instructors must not stereotype their students based on these characteristics, but rather treat each student as an individual.

If an instructor doubts that there is a range of FI–FD students, the student self-evaluation (Appendix A) may be given to the students along with or in place of the group embedded figures test (test booklets and a manual for administration are available from Consulting Psychologists Press, 3803 E. Bayshore Road, Palo Alto, California 94303). Chances are there will be a continuum of FI–FD learners in the classroom.

The instructor should also do the teacher self-assessment (Appendix A) to determine his or her own learning style. Since so many science teachers have been instructed in an FI manner, it is possible that they have automatically incorporated FI techniques into their instruction without careful thought as to why or if these techniques are really essential. Gross (1991) also indicates that one's own thinking style may influence one's approach to teaching more than one realizes. If this is the case, then instructors may very well instruct predominantly in the manner in which they prefer to learn.

As well, teachers need to be aware of their own biases in the classroom when grading, testing and calling on students. Research has shown that "people with similar perceptual styles tend to describe each other in highly positive terms, while people whose perceptual styles are different have a strong tendency to describe each other in negative terms. Field independence teachers perceive their FI students as being smarter than FDs, while FD teachers see their FD students as more intelligent" (Witkin et al., 1977). Renninger and Snyder (1983) also found that FD teachers perceived their FD students as learning more as did FI teachers with their FI students.

Classroom Activities

Within the classroom, teachers can better reach all of their students by guiding their classroom exercise format by what may be called *teaching to the poles*. Teaching to the poles refers to designing instruction to reach the students who are at the extremes of the continua of FI and FD styles. This approach helps ensure that more students will be reached, since the instructional method will include exercises for the entire range of FI-FD students (Felder & Silverman, 1988), which will by necessity also include those students in the middle of the continuum. Many instructors may already have an intuitive sense that there are different types of learners and to some extent may structure courses that have elements of FI and FD components, but without a clear sense of why they are doing it or how. By designing a course with the extremes of the FI-FD continuum in mind, instructors will offer the opportunity to succeed to both types of learners in a manner that is more clearly defined for both the student and the teacher.

Teaching to the poles entails dividing a course into sections and evaluating FI and FD instruction methods for each. For every classroom interaction, generally four to six FI-FD styles continua are most salient. (A full list of 15 styles continua are presented in the Cognitive Styles Assessment in Appendix A.) To determine the overall FI-FD nature of a course, the instructor needs to locate where each exercise sits on the continuum and determine if the course is heavily weighted toward FI or FD instruction.

As an example the authors have broken down a course into three areas: (i) communication of information, (ii) student information gathering and processing (laboratory and discussion) and (iii) evaluation of student learning.

Communication of Information

Since most introductory science courses necessitate conveyance of basic principles, a specific amount of lecture must occur. Since most people tend to have a style of lecture that is relatively fixed, the authors do not necessarily recommend that instructors change their lecture style. Within the context of lecturing, however, FD and FI strategies can be incorporated, especially if it is apparent that the overall structuring of one's lecture is more FD or FI.

Field independent teaching styles might include lecturing in a fairly impersonal manner, utilizing key words and equations on the chalkboard, and proceeding in a linear, sequential format. Lecture content would be fairly abstract, and relate strictly to scientific principles. The instructor would respond to questions, but most likely not ask the audience for answers.

Field dependent strategies for lecturing, on the other hand, might include a more personal lecture touched with humor, repeated use of slides, graphs, and chalkboard outlines. The lecturer would put the content in a social context, i.e., names of people who made major discoveries, or how a principle is currently in use in the modern world. For successful FD instruction, the professor could hand out a lecture outline or notes (advanced organizers) at the beginning of each class, or put a lecture outline on the board each day and leave it up for students to follow along. Or, the instructor might give the students a detailed course outline, rather than just a syllabus in the beginning of class. If possible, the instructor might engage students in the lecture, asking questions or asking the students to come up with the information themselves.

Student Information Gathering and Processing (Laboratory and Discussion)

Field independent work in the lab would be primarily individual and emphasize direct scientific principles. Laboratory work would probably be strictly quantitative and involve measurements and principles of basic science. The teaching assistant or professor would mostly lecture.

A more FD oriented laboratory and discussion would engage students in discussion, or have students do outside research and present information. Laboratory work would be group work, and experiments would be directly linked to subject matter in lecture as well as real life problems. Laboratories would also include observations as well as field trips to observe real-life structures and organizations.

Evaluating Student Learning

Field independent tests or problem sets would contain primarily quantitative, abstract problems. They would possibly be multiple choice, or short answer, fill-in-the-blank type of questions. The emphasis would be on offering questions that primarily have one right answer. Take-home work might be individual problem sets that would allow students some room to structure the problems. The entire course might be curved, to foster competition.

Field dependent assignments might consist of lab write-ups or research projects which would show how work is relevant to their own lives. Assignments could be given as group projects, tests would be long or short essay and the answers would be open to interpretation, or have a range of answers that could be considered correct. Quantitative problems would have a social context ("farmer Jill goes to check her field one morning and discovers some of her tomato leaves are green while others are yellow. She had applied 100 lb/ha of N as ammonium nitrate..."). The course or tests would not be curved, to foster cooperation.

EVALUATING YOUR OWN SYLLABUS FOR FIELD INDEPENDENCE–FIELD DEPENDENCE

Because each instructor has a different teaching style and offers various types of assignments, laboratories, discussion, and tests, a universal quantitative evaluation form for a syllabus would be extremely difficult to design. Table 5-1, however, models a syllabus for an introductory soil science course where each exercise is labeled FI or FD. Table 5-1 also shows one possible format of organization that instructors can use for evaluating their syllabi.

Table 5-1. Model Syllabus for Introduction to Soil Science.

Lecture

Lecture content/outline:

Section 1—Principles—first two-thirds of semester Introduction to Soils—what is a soil? Soil processes (genesis, formation, soil texture, etc.) Soil physical properties (water, volume, bulk density) Soil chemistry (ion exchange, liming) Soil biology (organic matter, major and minor nutrients, and microbial transformations) Soil fertility (integrate biology and chemistry) Soil management (integrate physics and processes)	FI†
Section 2—Contextual Problems and Applications—last one-third Comparative farming systems (conventional and organic) Waste disposal and soil contamination Site evaluations for development/preservation	FD-FI

Lecture style:

1. Give out syllabus as well as course outline along with syllabus in beginning of class, and refer to it each lecture	FD
2. Convey ideas first as abstract, then in context whenever possible	FD
3. Give handouts of graphs wherever possible so students have visual reinforcement to study from	FD
4. Assume large impersonal lecture	FI

Evaluation of Student Learning (60%)

1. Three full hour exams (at 15% of total grade each) with mixture of multiple choice, short answer, qualitative problem solving, short essay	FI
2. A. One optional final exam, more essay and contextual questions. OR	FD
B. 5-10 page original research paper using at least four original research references on applied environmental problems (effect of acid rain on soils, effect of tillage on soil erosion, etc.).	FI-FD

Laboratories and Discussion

Laboratory Exercises

1. Five observations and minor experiments Individual exercises (standard lab activities: testing pH, texture, bulk density, water movement)	FI
2. Three field trips compost station, farm, agency (USDA-SCS extension), soil observations (catenas, erosion, and genesis)	FD
3. Three two-week experiments with write-ups as group activities (testing fertilizers, microbial activity, and water infiltration)	FD-FI

Evaluation of Student Learning (40% for lab)

1. Three lab write-ups encompassing two labs each (2-4 pages with materials and methods section as well as results and discussions)	FI-FD
2. Three quizzes (short answer, some problem solving)	FI
3. One group presentation on preapproved subject	FI-FD

† FI = field independence; FD = field dependence.

For this model syllabus, the course was broken down into two sections, lecture, and laboratory and discussion. Both of these sections were then further divided into sub sections, the lecture into content and outline, style and evaluation; and the laboratory and discussion into exercises and evaluation.

Each of these subsections was then evaluated to determine if it had FI and FD components. For example, the *Principles* part of the lecture content and outline would be considered mostly FI, because it is abstract and non-contextual. The second part, the contextual part, is considered both FI and FD, FI because it involves the application of abstract ideas to concrete problems and FD because it is more global, and involves problems within a real life context. The laboratory exercise part has FI and FD components as well. The field trip and longer experiments are FD exercises because they entail contextual problems and group work. The shorter experiments are more FI because they entail individual work, and deal with more abstract ideas.

After labeling each subsection FI or FD, the author then counted the number of each to see if the course was balanced. In this case, the FI and FD parts of the course were weighted as they were tallied. For example, since the FI in the *Principles* section of the lecture content or outline part was for two-thirds of the semester, it would get weighted more heavily than the FD component of the *Contextual Problems and Applications* section, which was only one-third of the semester. The rest of the syllabus was then tallied in the same method.

Although this model syllabus offers one way of evaluating a syllabus or course for FI and FD, each instructor will of course have to make adjustments based on his or her own criteria. The two assessments of cognitive styles in Appendix A are meant to offer a guide to instructors as they develop their own criteria for evaluation.

While this model has offered some quantitative tools, instructors should also take a qualitative look at one's course and syllabus. Perhaps one of the most important considerations for evaluating a syllabus is to look at the self-assessments in Appendix A and the characteristics of FI and FD learners and then ask: does the course plan and evaluation method really allow for both types of learners to shine? Do the reading list and the lecture technique have both FI and FD components? Do both types of learners have an equal chance of getting an A? If an instructor can answer yes to these questions, and the numbers of FI and FD components in the course are relatively close, then chances are the entire class will have the potential to succeed.

CONCLUSION

We have tried to provide an overview of cognitive styles and offer some practical tools and models for adjusting classroom techniques to meet the needs of different types of learners. In soil science, university instructors are in a unique position because they may receive students who are still trying to decide whether to go into social science, or natural, physical, or biological science. Because cognitive styles are generally fixed by adolescence, soil science instructors will be facing many FD students who have been conditioned to believe that they probably cannot succeed at science, although it holds some interest for them. By expanding teaching styles, instructors can successfully teach to these students, and possibly bring in a whole new segment of people who will add to the richness and diversity of soil science.

The notion that science is only for a particular type of learner or thinker can no longer be used to justify the unidimensional manner of science instruction under the guise of tradition and quality. The world is changing too fast for that, and in the context of learning styles, it is very easy to see how this can be construed as an ethnocentric and gender biased argument. In biological sciences, diversity is considered necessary for the survival of ecosystems and the environment. Likewise, the field of soil science can only be enhanced by bringing in larger groups of people who will bring different ways of viewing the world, solving problems, and new ways of thinking.

ACKNOWLEDGMENT

The authors would like to acknowledge the instructors of the introductory soil science classes at the following universities for sending sample syllabi: Michigan State University, University of Wisconsin, Ohio State University, Washington State University, Oregon State University, University of Vermont, and North Carolina State. The authors also would like to thank F. Javier Martinez for assistance with research.

APPENDIX A

Even though the embedded figures test appears to be a valid and reliable instrument for determining cognitive style, two seemingly important criticisms in regard to its being used as the sole instrument for assessing cognitive style have lingered during its 40-yr life span. First, it really only measured field independence directly. If a subject can quickly disembed the simple figures from their complicated ground, he or she scores as highly field independent. Subjects who lack this ability are designated field dependent. Second, the performance of any group of subjects will array in a normal distribution, with most falling in the middle of the continuum, with a range from extreme field dependence to extreme field independence. Therefore the test is not agile enough to determine the specific perceptual or intellectual situations wherein those subjects in the middle range may prefer one cognitive style over the other (Ramirez & Castaneda, 1974).

In response to these criticisms, we have developed the *Cognitive Style Assessment* instrument for learners and teachers. This instrument is designed to directly assess preferences for both field dependence and field independence as well as assess specific situations where in subjects may apply alternate styles. The test may be administered alone, or, as the authors recommend, along with the embedded figures test.

Please note: this assessment is still in the experimental stages. If you use it in your classroom, we would appreciate feedback on how it worked and how you used it. Please contact Dr. Rodney Parrot, Office of Instructional Support, 14 East Avenue Cornell University, Ithaca, NY 14850.

Cognitive Style Assessment

Introduction

These instruments chart the cognitive styles of learners and teachers. They are primarily intended as tools for *self-assessment*, but, by altering the descriptions to emphasize behavior, could be easily adapted for *observation* also.

The instruments are designed to teach as well as assess. The 15 descriptions that form a column at the left of the dotted continua lines are all associated with field dependence, while those on the right are associated with field independence. Thus, at the same time that the student or teacher is determining his or her personal cognitive style, the instruments are introducing the test taker to the essential fruits of cognitive styles research in regard to field dependence–field independence as expressed in the educational enterprise.

The instruments are formatted such that, when completed, the array of bold **X** marks provide an at-a-glance impression of an individual teacher or learner's profile of cognitive styles. You may suggest that test takers use a felt-tipped pen to enhance this visual effect. Or the instruments can also be applied to groups of teachers or learners. Simply assign numbers 1, 2, 3, or 4 to the bubbles on the continua (left to right), tabulate and distribute on a curve for comparative study, or for statistical analysis.

Please note: These are suggested tools for cognitive styles assessment. Trials are now underway to investigate their reliability and validity. The authors would gladly entertain any suggestions for their improvement.

Directions for Use

Read descriptions at the left end and at the right end of the cognitive style continuum no. 1. Mark a bold **X** through the bubble that best locates you on the continuum, as a learner in *Assessing the Cognitive Style of Learners* or as a teacher in *Assessing the Cognitive Style of Teachers* on the following pages.

Mark only one bubble. Continue in the same manner for all 15 continua of cognitive styles.

Assessing the Cognitive Style of Learners

1. Prefer group interactive learning	<----- o ---- o --- | --- o ---- o ----->	1. Prefer solitary learning
2. Socially oriented in learning groups	<----- o ---- o --- | --- o ---- o ----->	2. Task oriented in learning groups
3. Prefer to cooperate for rewards	<----- o ---- o --- | --- o ---- o ----->	3. Prefer to compete for rewards
4. Very sensitive to criticism from others	<----- o ---- o --- | --- o ---- o ----->	4. Not influenced by criticism from others
5. Prefer informal, personal teacher	<----- o ---- o --- | --- o ---- o ----->	5. Prefer formal rofessional teacher
6. Very attentive to teacher's gestures	<----- o ---- o --- | --- o ---- o ----->	6. Very attentive to teacher's words
7. Prefer teacher to model material	<----- o ---- o --- | --- o ---- o ----->	7. Prefer teacher to explain material
8. Prefer to minimize distance to teacher	<----- o ---- o --- | --- o ---- o ----->	8. Prefer to maximize distance to teacher
9. Prefer ideas rooted in specific context	<----- o ---- o --- | --- o ---- o ----->	9. Prefer abstract ideas
10. Prefer to validate material	<----- o ---- o --- | --- o ---- o ----->	10. Prefer to critique material
11. Prefer structured learning exercises	<----- o ---- o --- | --- o ---- o ----->	11. Prefer to find own way to learn
12. Prefer ideas in context of a story	<----- o ---- o --- | --- o ---- o ----->	12. Prefer ideas in context of debate
13. Prefer narrative style writing	<----- o ---- o --- | --- o ---- o ----->	13. Prefer analytical style writing
14. Prefer holistic problem approach	<----- o ---- o --- | --- o ---- o ----->	14. Prefer sequential problem approach
15. Prefer to talk through problems	<----- o ---- o --- | --- o ---- o ----->	15. Prefer to think through problems

UNDERSTANDING COGNITIVE STYLES 43

Assessing the Cognitive Style of Teachers

1. Prefer facilitating group discussions	<----- o ---- o --- | --- o ---- o ----->	1. Prefer delivering lectures
2. Emphasize social orientation in groups	<----- o ---- o --- | --- o ---- o ----->	2. Emphasize task orientation in groups
3. Stress cooperation for rewards	<----- o ---- o --- | --- o ---- o ----->	3. Stress competition for rewards
4. Very sensitive to criticism by students	<----- o ---- o --- | --- o ---- o ----->	4. Very sensitive to self-criticism
5. Prefer informal, personal delivery	<----- o ---- o --- | --- o ---- o ----->	5. Prefer formal professional delivery
6. Very attentive to student's gestures	<----- o ---- o --- | --- o ---- o ----->	6. Very attentive to student's words
7. Prefer to model material	<----- o ---- o --- | --- o ---- o ----->	7. Prefer to explain material
8. Prefer to minimize professional distance	<----- o ---- o --- | --- o ---- o ----->	8. Prefer to maximize professional distance
9. Convey ideas as contexts	<----- o ---- o --- | --- o ---- o ----->	9. Convey ideas as abstract entities
10. Guide students to value material	<----- o ---- o --- | --- o ---- o ----->	10. Guide students to critique material
11. Assign structured learning exercises	<----- o ---- o --- | --- o ---- o ----->	11. Allow students to find own strategies
12. Convey ideas in context of a story	<----- o ---- o --- | --- o ---- o ----->	12. Convey ideas in context of debate
13. Stress narrative writing assignments	<----- o ---- o --- | --- o ---- o ----->	13. Stress analytical writing assignments
14. Encourage holistic problem approach	<----- o ---- o --- | --- o ---- o ----->	14. Encourage lienar problem approach
15. Favor students who talk through problems	<----- o ---- o --- | --- o ---- o ----->	15. Favor student who thinks for herself

REFERENCES

Anderson, J.A. 1988. Cognitive styles and multicultural populations. J. Teacher Educ. 39:2-9.
Anderson, J.A., and Adams, M. 1992. Acknowledging the learning styles of diverse student populations: Implications for instructional design. New Dir. Teach. Learn. 49:19-33.
Crow, L.W., and Piper, M.K. 1985. The effects of instructional aids on the achievement of community college students enrolled in a geology course. ERIC Document Repro. Serv. no. 256-566.
Davis, J.K. 1991. Educational implications of field dependence/independence. p. 149-176. *In* S. Wapner and J. Demick (ed.) Field dependence-independence: Cognitive style across the life span Lawrence Erlbaum Assoc., Hillsdale, NJ.
Demick, J. 1991. Organismic factors in field dependence-independence: Gender, personality, psychopathology. p. 209-224. *In* S. Wapner and J. Demick (ed.) Field dependence-independence. Lawrence Erlbaum Assoc., Hillsdale, NJ.
Felder, R.M., and Silverman, L.K. 1988. Learning and teaching styles in engineering education. Eng. Educ. 78:678-682.
Fennama, E., and J.M. Ayer. 1984. Women and education: Equity or equality. McCutchan Publ., Berkeley, CA.
Fields, S.C. 1985. Assessment of aptitude interactions for the most common science instructional strategies. ERIC Document Repro. Serv. no. ED 255 387.
Frank, B.M. 1984. Effect of field independence-dependence and study technique on learning from a lecture. Am. Educ. Res. J. 21(3):669-78.
Garlinger, D.K., and Frank, B.M. 1986. Teacher-student cognitive style and academic achievement: a review and mini-meta analysis. J. Classroom Instruc. 21(2):2-8.
Gross, R. 1991. Peak learning. Jeremy P. Tarcher, Inc., Los Angeles.
Mahlios, M.C. 1982. Effects of pair formation on the performance of student teachers. Action Teach. Educ. 4(2):65-70.
McDonald, E.R. 1984. The relationship of student and faculty field dependence/independence congruence to student achievement. Educ. Psychol. Meas. 44(3):725-31.
Moran, A.P. 1985 Unresolved issues in research on field dependence and field independence. Soc. Behav. Pers. 13:119-125.
Niaz, M. 1987. The role of cognitive factors in the teaching of science. Res. Sci. Technol. Educ. 5(1):7-16.
Ramirez, M. 1982. Cognitive styles and cultural diversity. ERIC Document Repro. Serv. ED 218 380.
Ramirez, M., and Castaneda, A. 1974. Cultural democracy, bicognitive development, and education. Academic Press, New York.
Renninger, K.A., and Snyder, S.S. 1983. Effects of cognitive style on perceived satisfaction and performance among students and teachers. J. Educ. Psychol. 75(5):668-676.
Riding, R.J., and D.J. Boardman. 1983. The relationship between sex and learning styles and graphicacy in 14-year old children. Educ. Rev. 35(1):69-79.
Scott, J. 1991. Lean on me. Summit Magazine, Winter, 1991-92:12-15.
Shade, B.J. 1984. Afro-American patterns of cognitive: A review of research. ERIC Document Repro. Serv. ED 244 025.
Sigel, I.E. 1991. The cognitive style construct: a conceptual analysis. p. 385-397. *In* S. Wapner and J. Demick (ed.) Field dependence-independence: Cognitive style across a life span Lawrence Erlbaum Assoc., Hillsdale, NJ.
Witkin, H.A. 1949. Perception of body position and of the position of the visual field. Psychol. Monogr. 63:1-46.
Witkin, H.A. 1952. Further studies of perception of the upright when the direction of force acting on the body is changed. J. Exp. Psychol. 43:9-20.
Witkin, H.A. 1978. Cognitive styles in personal and cultural adaptation. Clark Univ. Press, Worcester, MA.
Witkin, H.A., and S.E. Asch. 1948. Studies in space orientation: IV. Further experiments on perception of the upright with displaced visual fields. J. Exp. Psychol. 38:762-782.
Witkin, H.A., C.A. Moore, D.R. Goodenough, P.W. Cox. 1977. Field-dependent and field-independent cognitive styles and their educational implications. Rev. Educ. Res. 47:1-64.
Witkin, H.A., P.K. Oltman, E. Raskin, and S.A. Karp. 1971. A manual for the embedded figures tests. Consulting Psychologists Press, Palo Alto, CA.

6 Private Sector Experience of a Soil Science Graduate

Frances A. Reese

Larsen Engineers, Rochester, New York
State University College, Brockport, New York

ABSTRACT

This chapter focuses on the writer's experience as a graduate soil scientist working in the multidisciplinary environment of a consulting engineering firm. The utility of a soil science background will be related to professional experience in land use planning, environmental assessment, and solid waste management. Suggestions are offered to make soil science education more relevant to understanding complex environmental issues.

At the November, 1992 ASA-SSSA-CSSA-CMS meeting in Minneapolis, much discussion was focused on the topic of the future of the soil science profession. We heard how jobs with the traditional employers of soil science professionals, including government, academia and agribusiness, are diminishing. Government agencies are consolidating services and downsizing to accommodate tight budgets. University enrollment in traditional agricultural disciplines such as soil science are dropping. Businesses are downsizing and streamlining operations. Developments such as these affect professional employment opportunities for recent graduates and experienced soil scientists alike.

How do soil science professionals make the transition from government service and academic pursuits to the private sector? How do we educate (or reeducate) ourselves to compete in the changing job market? How does a soil scientist function in a nonagricultural business setting? As teachers, how do we help our students prepare for nontraditional careers?

Many soil scientists grapple with these questions at some point in their careers. I offer some ideas and potential answers to these questions.

BACKGROUND

This chapter is written from the perspective of one who is both a soil science educator and an environmental scientist in the private sector. I am

---

Copyright © 1994 Soil Science Society of America, 677 S. Segoe Rd., Madison, WI 53711, USA. *Soil Science Education: Philosophy and Perspectives.* SSSA Special Publication no. 37.

an adjunct instructor in the Earth Science Department at the State University College at Brockport, NY. The Department offers undergraduate degrees in geology, earth science, water resources, and meteorology. The course I teach is an introductory soil science class for upper level undergraduates and master's level graduate students. At the present time, the course catalogue lists no prerequisites for the course, although I generally discourage enrollment for students who have had no chemistry or biology coursework. Approximately two-thirds of the students are earth science, geology, biology, and meteorology majors, with the remainder being graduate education majors returning to college to obtain a certification to teach middle school and secondary level earth science, or, very infrequently, undergraduate education majors.

As an educator and as a project manager at a small environmental engineering consulting firm, I am aware of the importance of a sound scientific education for those considering careers in the environmental field. My multidisciplinary graduate studies in soil science and water resources, and an undergraduate degree in biology have given a good basic understanding of many complex environmental issues and problems. The discipline of soil science is unique because it incorporates knowledge from many other disciplines: biology, chemistry, physics, meteorology, and geology. If a student really wants to comprehend what is happening in the environment, he or she should study soil science. Serious study of several other scientific disciplines is required to understand the chemical reactions, biological activity and physical principles of soils thoroughly.

One disturbing trend I have noticed among students is that the general level of scientific understanding is declining. A second trend is the inability of many students to communicate adequately either in writing or orally. To combat these trends, students in my class are required to prepare a research paper using primary sources of information, and to present the paper orally at the end of the semester. Students are encouraged to write about any subject that interests them as long as it is somehow related to soil science. Many education majors and graduate teacher-students write lesson plans. Students are required to submit an abstract of their paper ≈ 3 wk into the semester. The abstract serves as a barometer about students' abilities, knowledge, and organization skills. From the research paper exercise, students learn bibliographic skills, organization, and technical writing. Because the oral presentations are short, they must be very focused and well organized.

Many students are intimidated by the prospect of having to write and present a paper in front of their peers. I have received many complaints about writing and presenting a paper in a science class. From my own experience, I respond that an individual cannot function in a professional employment environment without technical writing and oral communication skills. If an individual cannot communicate his or her results so that an employer, a reader, or a client can understand the significance of the work, the work itself is meaningless.

Soil scientists often have a difficult time being recognized as real scientists (Simonson, 1991). Simonson noted several instances in his career where his soil science background made him invaluable to his geologist colleagues. He also noted that the soil science discipline lacks the respect and recognition given to other disciplines such as chemistry and geology because soil science research has traditionally been empirical in nature and geared to the needs of the agricultural community. Soil scientists have not traditionally gotten involved in interdisciplinary activities or professional societies outside of soil science.

With the development of interdisciplinary societies such as the Society of Wetland Scientists, the American Planning Association, the Air and Waste Management Association, the Association for the Environmental Health of Soils, and the American Water Resources Association, soil scientists have more opportunities to become involved with disciplines outside the traditional agricultural realm of soil science. Two relatively new fields where soil scientists are finding employment are the fields of environmental assessment and site remediation. These fields are currently dominated by engineers and geologists; however, soil scientists are quite well equipped to tackle them. Field training in soil mapping, and coursework in remote sensing technology, air photo interpretation, and soil morphology, genesis, physics, and chemistry is especially useful in these fields. Most agencies require remediation plans to be prepared by a professionally licensed engineer; however, actual project design and management can be, and often is, done by others, including soil scientists. Soil scientists with experience in site assessment and remediation can be especially valuable in consulting activities because they often have more field experience and more training in chemistry and physics than many engineers. Firms with soil scientists on staff can compete very favorably for projects with consulting firms who only employ licensed engineers.

The growth of the number of private sector soil scientists in SSSA Division S-5 (Soil Genesis, Morphology, and Classification) has been documented by Miller and Brown (1987). These authors also note that "public sector soil scientists cannot possibly . . . meet all the day to day interpretive needs of clients." They also note the decreasing number of Ph.D.s in this field. A very recent paper by Boyle (1993) stresses the need for the soil science discipline to be revamped to meet the needs of the environmental industry. Boyle also comments on the tendency of the majority of soil scientists to stay within their traditionally funded roles.

Much debate has been centered on the need for professional certification in soil science. I support the national effort to certify soil scientists, but professional recognition of the credential has been slow, perhaps because there are so few soil scientists in comparison with the number of engineers and geologists.

I have not personally pursued American Registry of Certified Professionals in Agronomy, Crops, and Soils certification for one main reason. Experience requirements for the Certified Professional Soil Scientist (CPSS) credential seem to be heavily weighted toward individuals with academic or government service backgrounds. Individuals with soil science training working in the private sector find it difficult to meet the CPSS requirements without prior professional experience in government service or in an academic-extension setting. My 15-yr career did not include academic teaching until recently. My professional work experience in the private sector has not been strictly limited to the soil science discipline. In the private sector, particularly in the consulting field, one must function in many disciplines. My observation is that the government-academic experience requirement may be difficult for private sector soil scientists to achieve, unless they have had several years of prior government or academic service.

USE OF SOIL SCIENCE TRAINING IN THE PRIVATE SECTOR

My professional experience includes wetland delineation and mitigation studies, water quality studies, preparation of environmental impact analyses and statements, land use planning studies, natural resource inventories, development of local government ordinances and codes, permitting for solid waste facilities, and environmental site assessment for real estate transactions. Coursework in soil science was essential to many of the projects. This chapter summarizes three areas in which soil science training has been used: (i) wetland studies, (ii) siting and operation of a solid waste management facility, and (iii) environmental assessments for real estate transactions.

Wetlands Studies

Since the mid-1970s, the U.S. Army Corps of Engineers has regulated certain activities in waters of the USA under the authority of Section 404 of the Clean Water Act (U.S. Army Corps of Engineers, 1987). Wetlands are considered *waters of the United States*. I mapped federal jurisdictional wetlands using the *1987 Corps of Engineers Wetland Delineation Manual* as part of the design and environmental review of two highway projects located in Erie and Niagara Counties in Western New York.

Because both projects were federally funded, the project team was required to determine the location of both federal and state-regulated wetlands within the project area. New York State regulates wetlands 5.1 ha (12.4 acres) or larger. State-designated wetlands are mapped on U.S. Geological Survey quadrangle base maps using aerial photography and ground reconnaissance to verify boundaries. Under New York regulations, the presence of certain species or genera of vegetation is the primary criterion to determine whether an area is a state-designated freshwater wetland. The presence of hydric soils is not always required to designate a wetland under New York State regulations.

A three parameter approach is used to determine the presence of federal jurisdictional wetlands. Federal jurisdictional wetlands must show a predominance of hydrophytic vegetation, or be capable of supporting hydrophytic vegetation, and must show strong indications of hydric soils and wetland hydrology during the growing season.

One of my professional responsibilities is to manage the wetland service sector of our business. Expertise in assessing hydric soils, wetland hydrology, and field identification of plants is required to map wetlands efficiently, and to evaluate a site's mitigation potential. Engineers sought to minimize the impacts of the highway projects on wetlands in the project areas by shifting the highway alignment. Some wetland impacts and losses, however, were inevitable. The New York State Department of Transportation was required to mitigate wetland acreage losses.

I worked closely with the engineers to locate suitable mitigation areas within and adjacent to the project construction area, and to develop workable goals and objectives for the mitigation sites. An intimate knowledge of soil stratification, water movement, and the types of vegetation likely to survive in the proposed mitigation areas was required to accomplish this task. Soil test pits were dug throughout the proposed mitigation area, soil horizons were described, and periodic observations were made of water levels and movement patterns were made during a 2-yr period. Careful notes were made of vegetation species inhabiting the wetland area to be impacted. Observations were discussed and evaluated with agency staff from the U.S. Army Corps of Engineers and the New York State Department of Environmental Conservation. Mitigation goals and objectives were developed to guide the design process. Most of the area consisted of northern hardwood swamp and poorly drained, shrubby old agricultural fields. No open water habitat was available within the project area except for three small perennial streams. Because most of the habitat area was wooded, it was impractical to design an hectare for hectare, habitat replacement wetland mitigation area. Instead, the existing wetland values of wooded wetland habitat were enhanced by adding open water area, and creating a variety of water depths, shoreline and island areas, and emergent marsh complexes adjacent to the wet woods habitat. Approximately 2 ha of wetland mitigation area were created for every hectare of wetland lost to highway construction. Preliminary indications are that significant water quality and wildlife benefits are being provided by the mitigation area.

Siting and Operation of a Yard Waste Composting Facility

In 1988, New York's Solid Waste Management law (6 NYCRR Part 360) banned yard waste (grass clippings, wood chips, and leaves) from sanitary landfills. Communities were mandated to develop alternatives to recycle and reuse yard waste and other organic waste stream components by 1992. As a result, municipalities all across the state hastened to develop alternatives to landfilling yard waste materials.

My employer is currently designing or conducting preliminary tests on five compost sites in western New York. I am involved in several aspects of compost facility design, operation, and management. Graduate coursework in soil chemistry provided a basis for understanding the principles of organic matter decomposition and pesticide behavior in the soil–compost medium. This knowledge was applied to a problem experienced at one of the compost sites.

The site in question received yard waste materials from a variety of sources: private refuse haulers, golf courses, and nurseries. The site operator had little knowledge of or control over the application of pesticides and fertilizers to the materials to be composted. The composting site was located in an old gravel pit because it was in a sparsely settled, agricultural area and access to the site could be controlled. The site's owner was seeking an economic use for a property that was no longer suitable for either agriculture or mining. The site was located within an economical travel distance of the western suburbs of Rochester, NY. The site operator hoped to use the compost material as a soil amendment to reclaim mined out areas of the site, and other sand and gravel pits in the area. Negative siting factors included permeable sand and gravel subsoils. The depth to seasonal high water table was ≈ 1.3 to 1.9 m. The area selected for composting was located ≈ 608 m away from the nearest house, and ≈ 456 m away from a pit pond. No public water supplies were available to the area.

Local government officials expressed a concern about the impact that composting operations might have on a local pond and nearby private wells.

To address the concerns of local officials and residents, we monitored the nearby pond for ambient concentrations of commonly used pesticides, PCBs (polychlorinated biphenyls) and solvents. We had no funding available for groundwater monitoring. One of the realities of working with small businesses and entrepreneurs is having to work within very tight budget limits. In addition to budgetary restrictions, little information was available about the concentrations of pesticides and herbicides typically found in yard waste materials.

We obtained valuable assistance from a local landscaper and lawn service in addressing this question. We assembled information from the New York State Cooperative Extension and from our landscaper to develop a list of compounds that might be found in yard waste compost materials. We discovered the cost of laboratory analysis for these compounds prohibitive. From a literature review and inquiries to industry, I was aware of a field immunoassay test that might be used to detect the presence of low concentrations of a pesticide commonly used on lawns, 2,4-D (2,4-diphenoxyacetic acid). 2,4-D is commonly used to kill broadleaf weeds in lawns. We expected to find a residue of this compound in yard waste materials brought into the compost site.

We decided to use the immunoassay test to monitor compost product for the presence of 2,4-D and other related herbicides. We were faced with the problem of developing an appropriate method for extracting the target pesticide compound from a largely organic medium (compost), and developing reasonable dilution factors. This task was accomplished by working closely with industry personnel to adapt a method previously used on grains. Preliminary results indicated that raw yard waste compost contained measurable amounts of 2,4-D and related compounds. The field test used was not sensitive enough to determine actual compounds or exact concentrations. The method, however, was capable of detecting the family of compounds on an order-of-magnitude basis. We found this information useful for monitoring purposes. We tested the compost material when it was first brought in, at the mid-point of the composting process (≈ 4 wk), and at the end of the composting process. Our results showed that the concentration of 2,4-D and related compounds decreased rapidly during the decomposition process so that it was nondetectable in finished compost product.

This assignment required a knowledge of soil chemistry and biology. Soil science training provided the background to be able to ask the right questions and to find the resources needed to accomplish the task. Soils expertise was used to develop the list of compounds that might be found in the yard waste materials, to find an appropriate and affordable test method that could monitor the concentrations of a commonly applied pesticide through the composting process, to assist in the adaptation of the test method, and to explain the results to nonscientists involved in the decision making process.

Soil scientists have a niche in the field of solid waste management. Their expertise is best used in developing operational parameters for biological treatment systems, and in siting facilities.

Environmental Audits for Real Estate Transactions

During the late 1970s, Love Canal raised public consciousness about environmental pollution and contamination resulting from past or present misuse of property. The Comprehensive Environmental Response, Compensation and Liability Act (commonly known as CERCLA or Superfund) was enacted in 1980. This law made all current and former property owners, tenants, lessees, and financing institutions responsible for clean-up of contamination (Bureau of National Affairs, 1993). It created a firestorm of protest from banks and others who claimed they had no involvement or responsibility for creating environmental problems. The law quickly resulted in the need to develop a means of assessing a property's potential for environmental liabilities. Banks, attorneys, realtors, and property owners began requesting environmental audits of properties for potential environmental liabilities. Environmental audits are now required for all transactions involving industrial, commercial and multiple family residential real estate.

Environmental audits are designed to determine environmental liabilities that may be associated with real estate. The essential components of an environmental audit include review of the abstract of title; historic aerial photographs and maps; site plans; building plans and specifications; local, state, and federal environmental data bases and regulations; interviews with local government and regulatory agency officials; current and former owners, tenants, or lessees (where possible); and a site inspection, which may include environmental sampling.

Once the initial review and inspection are completed, the environmental audit report summarizes the scope of work, the resources and references used, the findings of the investigation, and details areas of potential or actual environmental liability. Further investigation and environmental sampling may be needed to characterize suspected problem areas, such as leaking underground storage tanks and piping or deteriorated storm sewers.

Environmental auditing became one of my project areas because it required many skills that I had already developed: air photo interpretation, the ability to use and understand environmental data bases and regulations, plan review and interpretation, and site assessment.

Environmental audits require a variety of skills and expertise. Complex environmental audits require a team of specialists, which may include environmental, mechanical and civil engineers, hydrogeologists, chemists, biologists, and soil scientists. I often function as a team leader and project manager on complex environmental audits because the job requires a generalist. My background in soil science has included training in chemistry, biology, and toxicology, disciplines that are not typically explored in depth in traditional engineering curricula. As a soil scientist working in a multidisciplinary environment, I have acquired a working knowledge of many disciplines. With years of experience, I have developed an understanding of many of the engineering facets. The project manager must be able to understand and integrate the information generated, to organize and direct the work efforts of project team members, and to communicate the results of the investigation to the client.

SUMMARY

Overall, I have found that soil science training provides a greater understanding of environmental problems and issues. It enables me to assist my engineering and scientific colleagues from other disciplines to develop practical solutions. The reason that soil science provides this advantage is its interdisciplinary nature. One cannot understand what happens in soil without having a good foundation in other scientific disciplines, such as biology, chemistry, physics, and meteorology. I do not regret my decision to study soil science, and in fact, would encourage others to do so. The respect problem outlined by Simonson (1991) is real, but one that can be addressed by improving our curriculum and teaching abilities, setting high academic standards for our students, and by getting actively involved in projects with other scientific and engineering disciplines.

REFERENCES

Boyle, M. 1993. Soil science. Environ. Sci. Technol. 27(5):813.

Bureau of National Affairs. 1993. Environmental due diligence guidelines. Updated monthly through January, 1993. Bureau of National Affairs, Washington, DC.

Miller, F.P., and R.B. Brown. 1987. Future developments in the private sector related to soil genesis, morphology, and classification. p. 269–278. In L.L. Boersma et al. (ed.) Future developments in soil science research. SSSA, Madison, WI.

Simonson, R.W. 1991. Soil science—Goals for the next 75 years. Soil Sci. 141:7–18.

U.S. Army Corps of Engineers. 1987. Wetlands delineation manual. Waterways Experiment Station, Vicksburg, MI. Technical Rep. Y-87-1. U.S. Army Corps of Engineers, Washington, DC.

7 Advising M.S. Graduate Students: Issues and Perspectives

Donald L. Sparks

University of Delaware
Newark, Delaware

ABSTRACT

One of the most important, satisfying, and challenging aspects of an academic position is the advising of graduate students. A very important part of graduate studies is the training of M.S. students, the primary focus of this chapter. I will discuss the following: advisor–advisee relationships, differences in advising M.S. and Ph.D. students, the role of the advisor in the student's research and professional development, and the importance of graduate student interactions with other graduate students, postdoctoral associates, and faculty.

One of the most satisfying, yet challenging aspects of a career in academia is the advising of graduate students. In my own case, there is nothing in my career that I have enjoyed more than advising graduate students and seeing them advance in their careers. Proper advisement is extremely important since the careers of future leaders in soil science will be greatly impacted by the quality of advisement they receive at the M.S. and Ph.D. levels. In Nielson (1970), Dan Hillel stated that the advisor should "encourage the development of a scientist and independent critical thinking rather than to teach the gospel sanctified because it happens to be the instructor's opinion." A very important part of graduate studies in colleges and universities is the training of M.S. students. Many of the points that will be discussed, however, are equally applicable to the training of Ph.D. students. I will discuss the following: advisor–advisee relationships, differences in advising M.S. and Ph.D. students, the role of the major professor in the research and the professional development of graduate students, and the importance of graduate student interactions with faculty, other graduate students, and postdoctoral associates.

Copyright © 1994 Soil Science Society of America, 677 S. Segoe Rd., Madison, WI 53711, USA. *Soil Science Education: Philosophy and Perspectives.* SSSA Special Publication no. 37.

ADVISOR–ADVISEE RELATIONSHIPS

Perhaps the most important aspect of graduate education is the type of relationship that exists between the advisor and the graduate student. For the relationship to be a good one, the advisor and the student must be compatible with each other. Compatibility will be enhanced if there is a careful and thoughtful system of selection and guidance of students that takes into account the abilities, personality traits, and expectations of the faculty member and student and matches each student to an advisor with whom there will be compatibility.

Moreover, the student must respect the advisor and the advisor should serve as a mentor to the student. A mentor can be defined as a close, trusted, and experienced counselor and guide. In a recent soil chemistry (Division S-2) newsletter, Dr. M.E. Sumner well stated the importance of mentors: "Take great care in selecting your mentors. They place an indelible stamp on you." As a mentor, one should act as a teacher to enhance the student's skills and intellectual development. Additionally, the mentor through his or her own personal achievements and reputation can serve as a person whom the student or protege can admire and emulate.

To be an effective mentor to graduate students, the advisor must lead by example and be competent and respected in the field, active and productive in research and in the profession, and familiar with the scientific literature. The advisor should also be industrious, self-disciplined, organized, creative, honest, enthusiastic, optimistic, humble, amicable, cooperative, patient, compassionate, and professional. I cannot overemphasize the importance of advisors being industrious, self-disciplined, and organized and the need for them to emphasize these characteristics to their advisees. These are truly necessary keys to a successful career and ones that advisees should possess. Advisors should stress to graduate students the need to be dedicated and to study and work long hours. Students should clearly understand that graduate studies are not just an 8 h a day job.

While it is important that students and advisors have a relationship that is characterized by mutual respect, compatibility, affability, and cordiality, it must be professional. Sorenson and Kagan (1967) found that many students desired a closer relationship with their advisors. Many desired to interact more with their advisors socially. For example, those students who were invited to their advisor's home felt that the professional relationship was enhanced and as a result, that they received better advisement and that there was greater progress made in their studies. I have seen some cases, however, in which advisors attempted to become one of the graduate student's peers and, consequently, the relationship and quality of advisement suffered.

Perhaps it would be instructive at this point to discuss what graduate students believe are the most important aspects of graduate student–advisor relationships and, particularly, graduate student advisement. There are few studies in the literature on this topic. Rugg and Norris (1975) conducted a survey of psychology graduate students on faculty supervision. Ten factors, listed below, were identified by the students as significantly affecting their satisfaction with the advisor. The advisor should be (i) flexible, fair, openmined, and supportive of student creativity and independence; (ii) provide structure and guidance; (iii) be a productive researcher and participate in research projects; (iv) have expertise in methodologies; (v) exhibit excellent interpersonal rapport by being friendly, relaxed, pleasant, and supportive; (vi) be a stimulating teacher both in the classroom and as an advisor of graduate students; (vii) be accessible to graduate students; (viii) be highly competent in one's field and be familiar with the current scientific literature, because this will facilitate discussions between the student and advisor on the research project and assist in interpreting the student's research findings; (ix) be mature and experienced in advising students; and (x) stress communications training for the student such as technical writing and public speaking courses, presentation of research results at meetings and conferences, and writing scholarly papers. The students surveyed felt that the latter was very important in their advancement and recognition.

In short, the findings of the survey by Rugg and Norris (1975) clearly indicate that graduate student advisors should not view their role as requiring little of their time, effort, or personal guidance. Rather, students want them to be actively involved in their overall graduate experience.

DIFFERENCES IN ADVISING M.S. AND Ph.D. STUDENTS

Many of the aspects of advising M.S. students are similar to those for advising Ph.D. students. There are, however, some fundamental differences. It is important at the M.S. level that students in soil science obtain excellent backgrounds in mathematics, chemistry, physics, microbiology, geology, statistics, and soil science, and that they become proficient in the use of computers for word processing and data analyses. A strong background in mathematics and the physical and biological sciences is particularly important if M.S. students in soil science plan to pursue Ph.D. degrees. For example, if a student wishes to pursue a Ph.D. in soil chemistry, he or she should take physical chemistry courses during his or her M.S. studies and not during the Ph.D. studies. Regardless of one's plans for the future, however, it is always useful to have fundamental training in mathematics, statistics, and the physical and biological sciences.

I feel that it is important that M.S. students have courses in quantitative and instrumental analyses. Moreover, they should become familiar with routine chemical, mineralogical, and physical methods for soil analyses through their research and coursework. I also believe that M.S. students should take courses in technical writing and public speaking since excellent oral and written communication skills are imperative for success.

Another aspect of advising M.S. students that differs from Ph.D. student advisement is the level of input into the research program and the degree of supervision by the advisor. Most M.S. students have not had significant experience in developing and conducting research projects. Therefore, it is imperative that the advisor be actively involved in the development of the research plan and meet with the student often about his or her findings, problems, and progress. I meet with my own M.S. and Ph.D. students at least every 2 wk to discuss their research and to exchange ideas. If this is not done, one may find that midway through the degree program, the student is floundering and no meaningful results have been obtained. The advisor should also ensure that the student is attempting to balance coursework and research. This is something that most M.S. students find different, since they have not had to manage their time to accommodate both coursework and research.

The degree of supervision at the M.S. level will also depend on the student's abilities. In the beginning of the research, more supervision may be needed; it can be reduced with time. The advisor should not spoon-feed M.S. students. If excessive supervision is given and an advisor's own ideas are imposed too much, a student's development is impeded and the student becomes a technician.

ROLE OF THE ADVISOR IN THE RESEARCH AND PROFESSIONAL DEVELOPMENT OF M.S. STUDENTS

The role of the advisor in the student's research is very important. The advisor should be in the forefront of his or her research area and expose the student to new information, hypotheses, and findings. Alexander (1970) notes that the student's work should focus on reSearch and not REsearch. The M.S. thesis should be original and not a reinvention of the wheel. To ensure this, the advisor should stress the importance of thoroughly reviewing the scientific literature and acquaint M.S. students with the major scientific journals. The student should be encouraged to read not only the contemporary literature, but also the older work in the field. Additionally, the advisor should recommend that the student read literature published in international journals, as well as those outside one's own field. A thorough literature review should be conducted and incorporated into the research proposal that is presented during the first semester or quarter of graduate studies.

In developing and deciding on the research project it is important to ask several questions (Bargar & Duncan, 1982): Is the research problem in concert with the student's developmental endeavors and creative capacity? Is the student excited and interested in the research problem? Will the research complement and broaden the student's abilities and insights?

At the outset of the M.S. student's studies the advisor should let the student know what his or her expectations are and encourage the student to communicate frequently. Such expectations should be the same for both foreign and U.S. students. The advisor should also make realistic and timely queries about the research progress. This lets the student know that the advisor is interested and also that he or she has expectations that progress be made. Advisors must be accessible to students and willing to talk and counsel with them. There should be regular meetings with students to review, evaluate, and discuss student progress. This will help in maintaining motivation and product-oriented behavior (Brown, 1968). The advisor should provide thoughtful criticisms of the research, both negative and positive, in a diplomatic and fair manner.

In turn, the student should strive to enhance his or her understanding of the research topic and methods that will be employed, analytically examine the research problem, critique his or her own work before seeking critical reactions from the advisor, and be responsive to criticism. If the student relies too much on the advisor, he or she could lose control of the research.

Thus, ownership of the research is important. It is shared between the student and the advisor, but the student must not lose control of the research. Ways to tell if ownership is being lost by the student include (Bargar & Duncan, 1982): (i) if the advisor discovers his own solution to a difficult part of the research and feels the solution is correct and that the student must accept the solution; (ii) if the advisor feels that the student has lost control of the research; and, (iii) if the advisor is more satisfied with his solution to a research problem than the student's solution.

The advisor should encourage the student to do the initial writing of the research, and then carefully provide input by meeting with the student and explaining point by point what is good and bad about the writing. The advisor should also have the student practice seminars and paper presentations, and stress the importance of publishing the research in a timely manner. The research should have a high degree of success and have a high probability of being published. The latter is important in advancing the student's career.

To enhance the success of the research, the advisor should be certain that there are financial and personnel resources and equipment available so that the student can carry out the research. The advisor should make sure that the student is exposed to modern equipment and learns new methodologies. It is a mistake, however, to allow a student's research to be overly dependent and structured around a piece of sophisticated equipment. Such instruments should be viewed as tools (Low, 1970). It is particularly important that adequate funds be available for laboratory, greenhouse, and field supplies, and travel to experimental sites and professional meetings. Advisors should ensure that M.S. students attend and present at least one paper at a professional meeting.

The advisor must also play an important role in the professional development of the graduate student. He or she should promote their advisees by nominating them for awards, introducing them to other scientists, assisting in job placement, providing information on job interviews, and stressing the importance of collegiality and image. While we usually do a fine job technically training graduate students, we do not spend enough time on developing our students professionally. One effective way that advisors and departments can assist graduate students in professional development is to offer a course that deals with grantsmanship, writing and reviewing manuscripts, resume preparation, job interviewing, and planning a work schedule.

IMPORTANCE OF INTERACTIONS WITH OTHER GRADUATE STUDENTS, POSTDOCTORAL ASSOCIATES, AND FACULTY

To enhance the experiences of M.S. students, it is important that they interact with other M.S. and Ph.D. students and postdoctoral associates both in and outside their research group. The advisor should also encourage M.S. students to discuss their research and their career goals with advisory committee members and with other faculty within and outside the student's department.

Students at the M.S. level in particular can benefit immensely from Ph.D. students and postdocs concerning methodologies and relevant scientific literature, and by engaging in discussions on their research. Also, the friendships and collegiality that are developed between fellow graduate students and others in the research group can be important throughout one's life. Thus, it is beneficial if the student's advisor has several graduate students of different academic levels. Having said this, I wish to point out that it is a mistake for advisors to have so many graduate students that they do not have time to advise each of them and provide input and guidance into their research and professional development. This is particularly important in advising M.S. students.

CONCLUSIONS

The future of soil science is bright. While there have been many important past successes in our field, numerous challenges and opportunities remain. These include: increased and more efficient food and fiber production, enhancement and preservation of environmental quality, and education of the public and elected officials about the importance of soil science. To be successful in these areas, we must have well-trained graduate students. The academic advisor is crucial in this regard.

DEDICATION

This paper is dedicated with admiration and appreciation to my outstanding graduate advisors, the late H.H. Bailey, who guided by M.S. studies; and D.C. Martens, and L.W. Zelazny who supervised my Ph.D. research.

REFERENCES

Alexander, M. 1970. Graduate instruction in soil microbiology. p. 35–41. *In* H.S. Jacobs and A.L. Page (ed.) Graduate instruction in soil science. ASA Spec. Publ. 17. ASA and SSSA, Madison, WI.

Bargar, R.R., and J.K. Duncan. 1982. Cultivating creative endeavor in doctoral research. J. Higher Educ. 53:1–31.

Brown, B.F. 1968. Education by appointment. Parker, West Nyack, NY.

Low, P.F. 1970. Graduate instruction in soil chemistry. p. 9-16. *In* H.S. Jacobs and A.L. Page (ed.) Graduate instruction in soil science. ASA Spec. Publ. 17. ASA and SSSA, Madison, WI.

Nielson, D.R. 1970. Graduate instruction in soil physics. p. 7. *In* H.S. Jacobs and A.L. Page (ed.) Graduate instruction in soil science. ASA Spec. Publ. 17. ASA and SSSA, Madison, WI.

Rugg, E.A., and R.C. Norris. 1975. Student ratings of individualized faculty supervision: Description and evaluation. Am. Educ. Res. J. 12:41-53.

Sorenson, G., and D. Kagan. 1967. Conflicts between doctoral candidates and their sponsors. J. Higher Educ. 38:19-24.

8 Supervision of Ph.D. Level Soil Science Graduate Students

Marion L. Jackson
University of Wisconsin
Madison, Wisconsin

ABSTRACT

The training of Ph.D. graduate students in soil science depends greatly on the proper selection of intellectually superior candidates and the identification of research topics suitable for a high degree of self-direction on the part of each student. Selection of a broad range of basic and applied study courses inside and outside of a given specialized field of soil science is also an important advisory function in the development of breadth. Paramount is the early and continuing development of a warm collegial and creative relationship between the teacher and the Ph.D. candidate in the out-of-doors, laboratory, office-conference room, and home. Collegiality should be used to smooth the potentially demoralizing transition of each student growing in personal life interdependent with growing professional responsibilities. The over-all aim of the supervision of Ph.D. soil science graduate students is the early promotion of the development of a strong personal and professional stature of maximum intellectual breadth.

Considerable interest has been expressed concerning the training of graduate students who become candidates for the Ph.D. degree in soil science. The process includes care in the selection of the candidate, selection of a research topic suitable for considerable self-direction, and the strategic selection of study courses. Success depends on the development of collegial and creative relationships that promote the development in the candidate of a strong personal and professional stature and intellectual breadth.

STUDENT SELECTION PROCESS

Success in advising Ph.D. graduate students begins with recruitment of candidates of superior qualities of intellect, personality, and personal drive. Obtaining the appropriate information will begin with the prospective student's application form that includes vital statistics and a single-page narra-

---

Copyright © 1994 Soil Science Society of America, 677 S. Segoe Rd., Madison, WI 53711, USA. *Soil Science Education: Philosophy and Perspectives.* SSSA Special Publication no. 37.

tive of his or her interests and qualifications. The applicant normally will arrange for one to three reference letters from qualified persons to be sent forward separately. This preliminary screening procedure greatly aids in the student selection process. Finally, the telephone often facilitates the recruitment of the best qualified prospective students.

In the early years, a young professor may be dependent on the reputation of his department and school to help attract good students. Very soon his own professional reputation built on research papers and student output will become the attractor, based on performance in the early years.

Occasional failure to make a favorable selection of a student will require weeding out later with attendant frustration and loss of time, money and self-esteem (Sorenson & Kagan, 1967). Besides student qualification difficulty, a poor interpersonal relationship between student and advisor often may be the cause of failure of the student. A success rate of Ph.D. completion of 90 to 95% can be expected if the selection process is effective, and the supervisor is competent in the subject matter, teaching expertise, and charisma (Rugg & Norris, 1975).

CHOICE OF RESEARCH TOPIC

The research topic will be selected through a matching of the advisor's interest and competence with the student's interest and career objective. To illustrate the breadth of opportunities offered to the soil science Ph.D. candidate, four examples are given:

1. Soil acidity and liming, the apparent dichotomy as earlier seen by E. Truog and R. Bradfield (H^+) and C.E. Marshall and H. Jenny (Al^{3+}), soil was seen as a proton donor through Al bonding (Jackson, 1963), $Al(OH_2)_6$ of $pK_1 = 5$, and quantum mechanical tunneling, a unifying concept of the soil acidity.
2. Soil adsorption of CO_2 by algal photosynthesis in soils has relevance to the role of CO_2 in possible global warming (committee on global change, 1988; Huang & Schnitzer, 1986; Arnold & Wilding, 1991; Kerr, 1992; Revkin, 1992) vs. the cooling effects of volcanic aerosols (Bryson, 1989).
3. Mineralogical analysis of a soil has scientific relevance only in the context of the soil landscape geomorphology over millions of years (Jackson, 1987).
4. Depletion of a soil-derived nutrient such as Se affects human health because an inadequacy of it in the food chain affects human longevity, heart disease rates, and cancer rates (Jackson, 1988).

In agricultural colleges in the USA there are frequently several sources of financial support for graduate assistants, some from the institution, federal agencies, and companies. Some agencies may be interested in having work done about the crop response to a product or its effect on the environment. The advisor tries to match the graduate student's interest to the agency interest. That matching often goes forward into the vocational selection that will be made after the Ph.D. is earned. Experience shows that research support from various sources can, with care, be successfully matched with student interests and background.

COURSE WORK

The advisor guides the soil science Ph.D. student's choice of a broad range of basic and applied academic courses, with the consideration of the student's interests, but also the requirements of science in general. Understanding of the soils is inherently a holistic and interdisciplinary challenge. But to provide for creative thinking, while avoiding the weakness of sensationalism, an adequate scientific grounding in course work is mandatory. Thus, the course selection must include basic subjects such as chemistry, physics, geology, climatology, plant physiology, microbiology, and mathematics as required to round out the course program already taken in undergraduate and M.S. programs.

The courses will cover a much wider subject-matter range than the thesis topic. Likewise, the scientific career will also cover a much broader spectrum of subject matter than the Ph.D. thesis. The Ph.D. thesis is thus only an early phase of a soil science career.

Extensive course work is an extremely efficient way of acquiring the needed breadth of learning through the accumulated experience of each of several instructors and course textbooks. Granted that a whole scientific career will be spent in learning, course work sets an efficiency paradigm for lifelong effective use of journals and books on a broad range of subjects. Primary and secondary school and college work have long proven their worth in accelerating the learning process. Courses taken in graduate school also have a large role in continuing the learning process. The Ph.D. stage is a fairly elaborate transition from formal schooling to self-teaching (Candy, 1991), not a quick-jump to it.

THE RESEARCH MEETING

A useful means of assisting the professional growth of the Ph.D. candidate is participating in research meetings with peers. A small group of 6 to 12 meet weekly at a set hour. Each one of the group can have a turn of leading the discussion of his or her research at a given weekly meeting. A one-page research report summarizes the long range objective and a second statement points out the objective of the immediate report being presented. A note on method, a few new data, and one or two references to related published literature round out the presentation. The next week a second student rotates through a similarly structured presentation. The research meetings thus markedly differ from a seminar series. Before the Ph.D. thesis is finished the student usually confides, "thank goodness for those weekly research reports."

As the home institutional work is progressing, regional and national professional meetings are attended and become increasingly meaningful in the growth of a peer network (Bargar & Mayo-Chamberlain, 1983). After a couple of years, the better Ph.D. candidate will be ready to offer a paper under the careful guidance and the help of the advisor. Thus, gradually the Ph.D. candidate will acquire a professional attitude and the formal requirements of the American Registry of Certified Professionals in Agronomy, Crops and Soils (ARCPACS) will be met (Bertramson, 1990).

CREATIVITY

The Ph.D. degree in soil science requires a segment of original research work performance, to bring out creativeness of the student. Essential reading in the library will then become increasingly more interesting, relevant, and significant. As Reason and Marshall (1987) have summarized, the advisor's intervention may foster growth in such categories as perspective, information, comfort, catalysis, and support. But in no case will the Ph.D. student expect to be told step-by-step what to do, beyond the initial days, as emphasized by Candy (1991).

It is important for the soils Ph.D. candidate to gain intellectual momentum by hands-on research in the laboratory and in the field. The supervisor can and should help these processes along. He will head a group of students out to examine instructive landscapes, to examine crops and soils in natural grassland and forest landscapes. Some samples will be brought back for laboratory analysis. Questions should be directed to the candidate, and some answers supplied as well. Once some momentum has been gained, the Ph.D. student may be surprised or even alarmed when items of specialized knowledge bring the student a step ahead of the advisor, which actually is to be expected. The doctoral thesis will reflect scholarship developed in these various ways (Bargar & Duncan, 1982).

PROFESSIONAL RELATIONSHIPS

The secret of success in advising Ph.D. students lies in the creation from the very first day onward of a warm interpersonal teacher–student relationship by one-to-one discussions in the out-of-doors, laboratory and office. Congeniality will be used to smooth the potentially demoralizing transition of the student growing into professional responsibilities (Bargar & Mayo-Chamberlain, 1983). The professor is anxious that the Ph.D. student show up well initially, and on to the final oral examination (defense of the thesis). To the extent possible, the advisor should be the student's sponsor or representative rather than an adversary. Having the student realize this helps to decrease the feeling of anxiety or panic that commonly threatens. A daily coffee break provides an opportunity for exchanging ideas between advisor, his student, and other students on an informal basis. A Saturday afternoon picnic or cook-out in the park or backyard at home does the same. On occasion, there will be a dinner at the professor's home. Departmental parties further extend the opportunity for congenial growth. A married graduate student family later on generally returns dinner invitations.

As the degree work progresses, the successful Ph.D. candidate and the advisor will have directed their thinking closely together. By graduation time, they will be peer colleagues, a relationship that will endure and grow with others at regional, national and international meetings. Not infrequently a coming together to continue studies or even for a sabbatic leave will ensue over the years. Former students' students (grandchildren, so to speak) also will become colleagues and friends. Society field trips will be enriched, as former students are in attendance—an extended-family effect.

REFERENCES

Arnold, R.W., and L.P. Wilding. 1991. The need to quantify special variability. p. 1–8. *In* Spatial variabilities of soils and landforms. SSSA Spec. Publ. 28. SSSA, Madison, WI.

Bargar, R.R., and J.K. Duncan. 1982. Cultivating creative endeavor in doctoral research. J. Higher Educ. 53(1):1–31.

Bargar, R.R., and J. Mayo-Chamberlain. 1983. Advisor and advisee issues in doctoral education. J. Higher Educ. 54(4):407–432.

Bertramson, B.R. 1990. Developing the professional mind-set. J. Agron. Educ. 19:191–193.

Bryson, R.A. 1989. Late Quaternary volcanic modulation of Milankovitch climate forcing. Theor. Appl. Climatol. 39:115–125.

Candy, P.C. 1991. Self-direction for lifelong learning. Jossey-Bass Publ., San Francisco.

Committee on Global Climate Change. 1988. Toward an understanding of global change. National Academy Press, Washington, DC.

Huang, P.M., and M. Schnitzer (ed.). 1986. Interactions of soil minerals with natural organics and microbes. SSSA Spec. Publ. 17. SSSA, Madison, WI.

Jackson, M.L. 1963. Aluminum bonding in soils: a unifying principle in soil science. Soil Sci. Soc. Am. Proc. 27:1–10.

Jackson, M.L. 1987. Roots of soil mineralogy—An introduction. p. 479–484. *In* L.L. Boersma et al. (ed.) Future developments in soil science research. SSSA, Madison, WI.

Jackson, M.L. 1988. Selenium: Geochemical distribution and associations with human heart and cancer death rates and longevity in China and the United States. p. 13–21. *In* G.N. Schrauzer (ed.) Proc. Int. Symposium on Present Status and Perspectives of Selenium in Biology and Medicine, Nonweiler, West Germany. May 1987. European Academy. Humana Press, Clifton, NJ.

Kerr, R.A. 1992. Fugitive carbon dioxide: It's not hiding in the ocean. Science (Washington, DC) 256:35.

Reason, P., and J. Marshall. 1987. Research as a personal process. p. 112–126. *In* D. Bond and V. Griffin (ed.) Appreciating adults learning: from the learner's prospective. Kogan Page, London.

Revin, A. 1992. Global warming. Abbeville Press, New York.

Rugg, E.A., and R.C. Norris. 1975. Student ratings of individualized faculty supervision: Description and evaluation. Am. Educ. Res. J. 12(1):41–53.

Sorenson, G., and D. Kagan. 1967. Conflicts between doctoral candidates and their sponsors: a contrast in expectations. J. Higher Educ. 38(1):17–27.

9 Advising Doctoral Students in Soil Science

Samuel J. Traina
*Ohio State University
Columbus, Ohio*

ABSTRACT

The Ph.D. represents the culmination of formal education for most research scientists in soil science. Thus it is imperative that the doctoral experience be positive, supportive and constructive. The interactions of Ph.D. candidates with their advisors can profoundly shape their perceptions of the scientific method, the scientific community and their future careers in soil science. Historically, much of this was conveyed to students through frequent contact with their advisors. Students were often allowed to develop at their own pace and could rely on their advisors for extensive guidance. Unfortunately the growing demands for increased extramural funding, the need to meet contract deadlines, and the national trends of increased teaching workloads for university faculty are potentially damaging to this relationship. Nevertheless, Ph.D. advisors must recognize the developmental aspects of the doctoral experience and provide for sufficient time for the establishment of constructive advisor-advisee relationships with their graduate students.

The Ph.D. degree represents the culmination of formal education in soil science. As such, the doctoral experience can have a profound influence on the professional and personal development of students. We have long recognized that the choice of course work and research topics can set the foundations for the technical skills on which an individual builds his or her career, but how much attention do we pay to the topics of scientific creativity and freedom, intellectual risk, and ownership of research ideas? What is the extent to which these issues impact on our role as advisors of Ph.D. students in soil science? What influence is the changing role of agriculture and soil science in academia and our society having on these interactions? I will attempt to discuss these topics in the context of the formal education literature and from my own personal perspectives.

In a treatise on the cultivation of creative endeavor in doctoral research, Bargar and Duncan (1982), indicated that formal discussions and presentations of research in texts, articles, and papers do not convey the true nature

Copyright © 1994 Soil Science Society of America, 677 S. Segoe Rd., Madison, WI 53711, USA. *Soil Science Education: Philosophy and Perspectives.* SSSA Special Publication no. 37.

of scientific inquiry. Such literature tends to portray scientific creativity as a linear progression of logical thought based on careful extrapolation of existing knowledge. This is particularly true when one considers the development of a hypothesis. Yet as these authors point out, there is no discussion of the psychological means of how a sound hypothesis is created. Where do *good* hypotheses come from? How do we account for their originality, and their ability to stimulate the creative mind. How do we decide which hypotheses to work on? Traditional dogma would suggest that all of these issues are dealt with by linear, logical thinking, but this may not always be the case. Biographies of scientists, deemed creative and eminent by their peers often contain accounts of nonrational or intuitive insights into a given problem (Bargar & Duncan, 1982). This intuitive thinking leads in turn to the development of an apparently logical hypothesis. Kekule discovered the structure of the benzene ring during a dream of a serpent biting it's own tail. Whereas, this is perhaps the most renowned example of intuitive scientific thinking, it is by no means unique. It is not suggested that intuition and nonrational insight should replace the formal deductive approach to scientific inquiry; but rather that it is a critical part of the entire process of "creating new scientific thought." A rigorous, ratioale development of linear logic may best serve to test, evaluate, clarify, and amplify a hypothesis. But creative intuition is still required to guide investigators in their search for knowledge. As Bargar and Duncan (1982) point out

> With emergent scientific insight often comes a sense of excitement and a valuing of the meaning and potential implications of the insight. This excitement and sense of value helps stimulate the commitment and dedication necessary to fruitful and sustained creative work.

It is our task as advisors to facilitate the development of this insight and creative excitement in our students. This is a formidable task and there are no clear cut and simple ways in which to accomplish it. Yet the advisor can do much to create an environment conducive to creativity. Graduate students must be strongly encouraged to delve into the scientific literature. These excursions should not be limited to narrow confines of their advisor's discipline, but should extend to many different areas in the physical and natural sciences. Often some of the most creative ideas involve transferring an approach or logic structure from one discipline to another. Concomitantly, doctoral students should be strongly encouraged to attend seminars and discussion groups in a broad range of subject areas. Exposure to many different concepts and perspectives can do much to stimulate the creative process.

Change itself, can have a profound influence of the ability of a doctoral student to act creatively. Bargar and Mayo-Chamberlain (1983) stress that "entering a graduate program often involves considerable dislocation of personal life, including a geographic move and a lowering of income. The person's daily activities and schedule may have changed drastically, and the psychological environment may be sufficiently different to generate culture shock. It is natural for these conditions to prompt anxiety and raise doubts about whether the change in life-style is worth it. Students often feel at sea: Challenged and determined on the one hand and uncertain and anxious on the other." They look to us as advisors for professional and academic guidance and encouragement. It is our responsibility to provide the proper environment conducive to the development of creative, intuitive thinking, while at the same time allowing each individual the time to find their own equilibrium in their new environment.

The first point at which the doctoral advisor can aid in a student's education is in the development of his or her course program. Whereas, it can be assumed that all graduate students in soil science require a strong background in the fundamentals of mathematics, statistics, and the physical and natural sciences, specific subject areas of emphasis should be chosen jointly by the student and the advisor. Barrage and Mayo-Chamberlain (1983) feel that "creative individuals see themselves as having the authority to behave with independence, to venture beyond the accepted into new terrain, and to be responsible for the outcome. They must summon the courage to act on that vision." Such a vision is particularly relevant to the rapid changes occurring in the discipline of soil science. In the last few years, we have seen a diminishing of the perspective of soil science as a subdiscipline of agriculture. In its place we have begun to recognize the significant contributions that soil science can make to the fields of ecology, environmental science, land use, and many other disciplines. This warrants a careful examination of each course program, to insure that it meets the needs and aspirations of both the advisor and the student. It is important to recognize that while an advisor may be concerned about the image projected by their students, and the impact that their courses may have on their immediate research activity, it is the student that must live with the consequences of the course choices throughout his or her career. They must be sufficiently vested in basic knowledge to allow them to accommodate changes in research focus throughout their lifetimes, but they must also be adequately educated to enter the work force upon completion of their degree. Overall, it is crucial that the advisor and the advisee arrive at a mutually shared perception of what represents an acceptable course program. Enrollment in classes offered by a variety of other disciplines, should not only be encouraged, it should be required. In the present climate of restricted university budgets and declining enrollments this may seem at odds with departmental needs; nevertheless, it is critical that we provide the greatest opportunity to foster creative thinking in our graduate students. This necessitates that they extend their formal education beyond the confines of soil science, and that they are readily able to communicate and interact with a broad scientific and public community.

Perhaps the most important aspect in developing creative thought processes in dissertation research is the selection of a research topic. Students come to this point through a number of different paths. Some enter graduate school with a clear and well charted research topic in mind, but most simply know that they wish to pursue graduate studies in a general subject area such as soil genesis. Choice of a dissertation topic should begin early in the degree program so that the student's classes and readings may contribute as much as possible to the development of a viable hypothesis. Topic development can be viewed as a problem solving process in which advisors take a facilitating role in helping students articulate and assess alternatives. The advisor should insure that the student chooses a research topic that is rigorous and warrants investigation. It is also highly desirable that consideration be given to what future employment opportunities may be available to the student upon completion of his or her research. At the same time, the advisor must maintain a degree of distance from the thesis topic to avoid excessive influence on the student's choice. Bargar and Duncan(1982) suggest that this can be accomplished by:

1. encouraging the student to talk openly about all presently relevant aspects of the research endeavor;
2. listening thoughtfully to the student's accounts;
3. explicating at appropriate times the advisor's understanding of what the student has said; and
4. asking the student to confirm or disconfirm the advisor's understanding of the student's present view of the creative research endeavor.

This model clearly suggests that the choice of a research topic should be made after careful consideration by the student and his or her advisor, but that the primary choice of subject area should be made by the doctoral student. Such an approach represents an ideal model of doctoral education. Unfortunately, there are many pressures that make complete adoption of such an approach quite difficult.

One such pressure is funding. Since about 1985, agricultural research in general, and soil science research specifically, has experienced some dramatic changes. Formula funded research supported by Agricultural Experiment Station moneys has greatly diminished in many academic institutions. To some extent, this has led to a decline in agriculturally oriented soil science research. Perhaps more importantly, American universities nation-wide are facing moderate to severe budget cuts and financial set-backs. In many institutions this has led to a reduction in institutional graduate student support. Many departments that conduct graduate education in soil science have less dollars for graduate student assistantships. Concomitant with these reductions in institution dollars, has been an increase in the number of opportunities for extra-mural research in soil science. Much of this research is driven by concerns about land use and the environment. Issues such as nuclear and hazardous waste disposal, surface and ground water contamination by industrial and agricultural chemicals, global climatic change, and sustainable agriculture are all taking soil science in new directions. These new funding opportunities can open many doors for us as research scientists that allow us to broaden the perspectives and experiences of our students, and provide them with research stipends; but what do they do to our student's scientific creativity? If we as principle investigators write grant proposals with clearly defined specific hypotheses, can we let our students conduct these projects as their dissertation research? If so, what has happened to the student's input in hypothesis development, and how are we as advisors fostering their creative thinking. One answer to this potential dilemma would be to prepare grant proposals cooperatively with graduate students, however this is often difficult due to logistical constrains (funding and time). Clearly the changes in funding sources will continue to change the research that we conduct as soil scientists, which in turn may change the creative freedom that we can allow our students to have in choosing their research topics. The challenge then is to maintain sufficient input from the student, so that the doctoral research truly contains their creative ideas, while addressing the needs and goals specified in a research contract.

On the topic of conducting the dissertation research, Bargar and Mayo-Chamberlain (1983) indicate that "advisors can be of genuine assistance during the research process with a variety of activities ranging from the practical aspects of research methodology to the more subtle aspects of synthesis and critical review." The stimulation of students to arrive at their own critical thinking and synthesis is stressed. This ideal model must also be evaluated in light of the changes in our discipline. How much freedom can we give a student that is working on an external sponsored research project for his or her dissertation research? Graduate students must be allowed to make mistakes and find their own solutions to problems in their research. They must be encouraged to take risks and to explore new directions. This can cause an enormous amount of self doubt and introspection. If they are working on a truly new area, doctoral students can feel as though they are standing alone, on the edge of a frontier, possibly in opposition to established scientific dogma. On these points, advisors must encourage their students to follow their visions, while simultaneously guarding against truly wrong directions. This can be a timely process. "Creative individuals, cannot command, cajole, or force their own minds to be productive but must in fact learn to cooperate with processes and forces that move by their own timing and not at the will of the conscious ego," (Bargar & Duncan, 1982) or in the case of graduate students, at the will of their advisors. Yet, to insure continuous and future funding the graduate advisors must file research reports to external funding agencies in a timely manner. They must conduct good and judicious research. In the absence of external funding, inadequate research results can remain internal to the educational institution. In the worst cases a graduate student may not complete his or her degree. While such incidents are clearly tragic and undesirable they generally have stronger impacts on the advisee than the advisor. If a given research activity is funded by an external agency, however, the doctoral advisor will probably bare primary responsibility for its outcome. Principal investigators generally cannot abdicate responsibility for research conducted under their guidance by blaming it on an inadequate student. So we have a quandary. Students need to be able to make mistakes, but advisors need error free, rapid research. This is an issue that is not readily resolved, and is likely to be a growing problem in our discipline as we continue to shift away from *formula funded* to *competitive* research activities. As advisors, we must approach this issue with open and directed caution.

Ownership of research ideas is a topic related to graduate student freedom and creativity. "For all practical purposes, in doctoral work, ownership is shared; but students must be given every reasonable opportunity to take responsibility for a problem and its solution" (Bargar & Mayo-Chamberlain, 1983). How is this readily accomplished if the student's research topic is part of a grant proposal written by his or her advisor? Does not the advisor have vested interest in the ownership and outcome of the research. Under these conditions, it is possible for an advisor to discover their own solutions to some aspect of the student's research problem and to believe that their solutions are *the* correct approach, as opposed to some solution that the student may propose. In such instances, the advisor has assumed intellectual ownership of the student's research. This can lead to feelings of intellectual inadequacy in the student and possibly future conflicts with the advisor. Ownership of research concepts is a difficult issue, perhaps best described by Bargar and Mayo-Chamberlain (1983), "Given the dynamics of the mentor-mentee relationship in advising, the problem of *ownership* can be subtle and potentially troublesome."

What then is the ideal role of the professor in advising doctoral students in soil science? The advisor must serve as both a resource and a role model for the student. In the former case, the advisor makes available to his or her students, their previous experience, their research methodologies and expertise and their perspectives on the future of the discipline. In doing so they must maintain a strong interest and involvement in the student's research, while being distant enough to provide objective evaluation. In the later case the advisor can share with the student their excitement and enthusiasm for science, and the ways in which they balance their careers and their personal life. Additionally, the advisor must play an ever increasing role as grantsman and employer, obtaining external research dollars to support the financial needs of the student. As discussed above, this can provide additional sources of strain for the advisor-advisee relationship, regarding issues of timeliness and quality of the student's research activities. In some instances, it can also reduce the advisors role to one of a fund-raiser, rather than an educator. Ideally, the advisor maintains a balance between research director, intellectual guide and confidant, and employer, keeping both his or her interests and those of the graduate student in focus.

Perhaps the greatest contributions that we can make to our profession as soil scientists is the successful education of doctoral students as soil scientists. The development of human potential can far exceed the impact that we might make through the development a new postulate or theorem. It is our responsibility as educators to insure that this is a positive and supportive process. We must give our students the freedom to develop their own scientific creativity, and to stumble through the pitfalls and setbacks at their own pace, while at the same time recognizing our commitment to funding agencies and to society as a whole to produce quality scientific information in a timely fashion. This can be an arduous task, but the rewards far exceed the effort.

REFERENCES

Bargar, R.R., and J.K. Duncan. 1982. Cultivating creative endeavor in doctoral research. J. Higher Educ. 53:1-31.

Bargar, R.R., and J. Mayo-Chamberlain. 1983. Advisor advisee issues in doctoral education. J. Higher Educ. 54:408-432.

10 The Advisor–Advisee Relationship in Soil Science Graduate Education: Survey and Analysis

Philippe Baveye and Françoise Vermeylen
Cornell University
Ithaca, New York

ABSTRACT

Two different questionnaires were sent separately to soil science graduate students at U.S. and Canadian universities, and to their faculty advisors. Among other things, these questionnaires were meant to serve as a basis for an analysis of various aspects of the advisor–advisee relationship. This analysis revealed a number of areas that are prone to misunderstandings and miscommunication. They include the level of directiveness of the advisor, the preparation of the students for their future career, and the difficult issue of the ownership of the research output. In spite of frequent divergences of perception on these points, advisees and advisors nevertheless, expressed views on each other's performance, that were generally positive.

For many graduate students, the quality of the *human* relationship they have with their advisor is an essential component of their M.S. or Ph.D. program. Sorenson and Kagan (1967) considered it so essential that they suggested that instead of selecting among applicants for graduate studies solely on the basis of academic attainment, "what is needed instead is a system of selection and guidance that takes into account the abilities, personality traits and expectations of faculty members and students, and matches each student to a sponsor with whom he or she will be compatible."

Without going necessarily as far as this psychological *match-making*, it seems important, for the graduate experience to be successful, to make sure that the advisor–advisee relationship be positive, open, frank, and supportive. Of course, as in any relationship, one expects that the student and his or her advisor will not always see eye-to-eye on everything. In some cases, it may take several or even many years for the student to understand his or her advisor's viewpoint on certain issues. Nevertheless, it seems important for both parties involved to be willing to spend the amount of time needed to establish a good dialogue, so that each knows precisely where the other

---

Copyright © 1994 Soil Science Society of America, 677 S. Segoe Rd., Madison, WI 53711, USA. *Soil Science Education: Philosophy and Perspectives*. SSSA Special Publication no. 37.

stands. Concerted efforts should be made by both parties to quickly resolve any misunderstanding or miscommunication that may, and often does, arise during the course of M.S. or Ph.D. programs. Diagnosing these gaps in perception is not always straightforward, however.

The primary objective of the research described in the present chapter was to identify the aspects of the advisor–advisee relationship that are most prone to misunderstandings or communication gaps. This required us to solicit input both from students and from their advisors, an exercise that apparently had never been done previously in this particular context. It is hoped that the research findings reported in the next few pages will be of interest and of some help to soil science graduate students and to their advisors.

MATERIALS AND METHODS

The objective outlined in the introduction above could have been reached in a number of ways. We decided to have recourse to survey instruments. They suffer from a number of drawbacks (e.g., Wiersma, 1969; Best, 1970), not the least of which is the necessary assumption that the individuals surveyed have a sufficient grasp of English to understand the survey questions (several faculty supervisors argued that it was not true of their advisee). Survey instruments, however, have the definite advantage that they allow investigators to work with large samples, representative of the whole population. To minimize some of the very real difficulties associated with the design of these survey instruments, we chose to modify an existing instrument, conceived and thoroughly tested to meet objectives similar to ours, even though its authors (Rugg & Norris, 1975) were concerned only with students' perceptions.

Survey Instruments

The 51 item supervisor rating instrument developed by Rugg and Norris (1975) was used as a starting point in the elaboration of two survey instruments appropriate for the purposes of this study. Some of the items in Rugg and Norris' (1975) instrument were slightly modified whereas others were entirely eliminated. Also, a number of new items were added to cover specific aspects not addressed in Rugg and Norris' (1975) survey. The students' survey instrument was pretested within the Department of Soil, Crop, and Atmospheric Sciences at Cornell. Various comments and suggestions resulting from this pretesting were taken into account in revising the initial instrument format.

The final survey instruments that were sent to the 300 students and to their faculty supervisors consisted of three parts: (i) a list of 33 statements on various aspects of the supervisor-student relationship; (ii) a series of statements on the level of satisfaction of the supervisor with the student or, for the student, with the faculty supervisor, and (iii) two questions concerning the major function of the faculty advisor.

In the first part of the survey instrument sent to the faculty supervisors, the 33 statements, listed in sequence from 1 to 33, were as follows (the percentages after each question will be referred to in a later section of this chapter).

RS: Respect for students

2. You encourage independent work on the part of the student. 3.9%
8. You decide in detail what is to be done in the research and how it is to be done. 30.9%
20. You have difficulty communicating in meetings with the student. 16%
22. In publications and talks, you take personal credit for the student's work. 22.9%
23. You have little confidence in the student's ability and integrity. 11.1%

SG: Structure and Guidance

4. You give appropriate constructive criticism of the student's work. 16%
6. You help to relate the student's project to the student's short term and long term goals. 19%
12. You help to clarify specific objectives to be met by the student during the research. 16.4%
16. You are willing to recognize the limits of your knowledge, expertise. 16.3%
18. You direct the student to other relevant resources or individuals with expert knowledge. 13.3%
24. You are uninterested in and unenthusiastic about the student's project. 10.4%
26. You recognize work well done by the student. 11.2%
29. You are concerned about the overall value of the experience for the student. 18.4%
30. You schedule consultation-progress report meetings with the student. 19.3%
31. You involve the student in the entire research process, from writing the proposal to publishing the results. 10.7%

RP: Research Productivity

5. You are actively engaged in research. 11.6%
28. You frequently submit articles and manuscripts for publication. 9.1%

RM: Research Methods Expertise

15. You are very familiar with research design principles. 13.3%

IR: Interpersonal Rapport

    10. You treat the student like a collaborative colleague.   17.2%

    14. You pay attention to other aspects of the student's life besides the student's studies.   28.7%

ST: Stimulating Teaching

    9. You enjoy supervising graduate students.   14%

    19. You are familiar with the content of a wide variety of specialty areas and related fields.   10.2%

    25. You confront the student with alternate procedures, interpretations, and ways of expressing ideas.   22.8%

SA: Supervisor Accessibility

    7. You often have difficulties, because of other constraints, to schedule time for meetings with the student.   10.1%

    21. You are willing to provide help when the student needs it. 8.8%

SM: Subject Matter Expertise

    1. You demonstrate comprehensive knowledge on topics of interest to the student.   11.6%

    11. You are familiar with current developments in the student's field of interest.   13.2%

    17. You have professional and research interests that overlap with the student's.   13.2%

CT: Communications Training

    3. You help the student to give better lectures/seminars.   15.3%

    27. You provide helpful critiques of the student's writing style.   14.2%

CP: Career Preparation

    13. You help the student to make contacts that could be useful for the student's career.   27.3%

    32. You are concerned about preparing the student for all aspects of the student's future career, not just for the research   29.84%

    33. You are willing to provide financial support so that the student can attend scientific conferences.   22.1%

The headings in this list were not included in the survey instruments. They correspond to nine of the 10 first-order orthogonal factors identified by Rugg and Norris (1975), supplemented by a new factor labeled Career Preparation. The Faculty Maturity factor of Rugg and Norris (1975) was not used. In the research described in the present chapter, no attempt was made to carry out a factorial analysis similar to that of Rugg and Norris

LEVEL OF SATISFACTION

Item #	How satisfactory do you find:	Very unsatisfactory Very satisfactory
34	the overall performance of the student	☐ ☐ ☐ ☐ ☐ ☐ ☐
35	progress made to date by the student toward his/her research objectives	☐ ☐ ☐ ☐ ☐ ☐ ☐
36	the student as learner	☐ ☐ ☐ ☐ ☐ ☐ ☐
37	level of independence of the student	☐ ☐ ☐ ☐ ☐ ☐ ☐
38	your social relation with the student	☐ ☐ ☐ ☐ ☐ ☐ ☐
39	the response of the student to your supervision	☐ ☐ ☐ ☐ ☐ ☐ ☐

Fig. 10-1. Second part of the survey instrument sent to the faculty advisors.

(1975). The headings should, therefore, not be looked at, at this stage, as more than a convenient, but very approximate, way of classifying the survey questions.

In the survey instrument sent to the graduate students, the 33 statements above were suitably modified to address the students' viewpoint, without however changing their meaning in the least. For example, the first statement (no. 2) in the list above became: "Your faculty supervisor encourages independent work on your part".

The students and their faculty supervisors were asked to react to the 33 statements via a seven-point Likert scale ranging from no/almost never to yes/almost always. As in the survey instrument of Rugg and Norris (1975), there was no systematic arrangement of the statements, aside from the fact that negatively worded statements (e.g., no. 24) or statements normally inviting negative responses (no. 20) were interspersed with positively worded items to discourage the respondents from adopting automatic response patterns.

The second part of both survey instruments was a short series of statements concerning the level of satisfaction of the student with the faculty supervisor, or vice versa. In the survey instrument sent to the advisors, there were six such statements (Fig. 10-1). The student's survey instrument, on the other hand, had nine statements directly inspired by the work of Rugg and Norris (1975). They addressed such issues as faculty supervision (no. 34), subject matter (content) learned (no. 35), supervisor's interpersonal style (no. 36), overall value of the experience (i.e., your current degree program) (no. 37), supervisor's subject matter expertise (no. 38), research skills learned (no. 39), supervisor as teacher (not only in classroom!) (no. 40), progress toward your initial goal for the experience (no. 41), supervisor's research method expertise (no. 42). In both survey instruments, seven-point Likert scales were used.

GENERAL PERCEPTION

According to Brown (1968), the faculty supervisor has three major functions:

(1) to identify references and resource individuals appropriate to the student's project concerns

(2) to provide constructive feedback to the student on the plan of action developed to achieve project goals

(3) to meet regularly with the student to review, evaluate, and discuss student progress in order to maintain motivation and "product-oriented" behavior.

Which of these functions would you rank as most important?

(1) ☐   (2) ☐   or   (3) ☐

Which one would you rank second?

(1) ☐   (2) ☐   or   (3) ☐

Fig. 10-2. Third part of the survey instruments, sent both to the advisees and to their advisors.

The third part of both survey instruments (Fig. 10-2) consisted of two questions concerning the major functions of the faculty supervisor, as defined by Brown (1968).

Before mailing them, both survey instruments were identified by three-digit numbers, which allowed us eventually to match each student's responses to those of his or her advisor. In the cover letter we sent with the survey instruments, we assured students and faculty advisors that their identity would never be revealed to anyone, in other words that our analysis of the data and our reporting of the results would be strictly name-blind.

Students List

Sixty-six institutions in the USA and Canada currently offer graduate degrees in soil science. A request was made to all of them, in early 1992, for a list of their M.S. and Ph.D. students in this field, along with information on these student's gender, nationality, and degree program, as well as on the name of their advisor. All institutions except five responded to this initial survey, providing data on a total of 1280 students. Summary statistics for this preliminary part of the research are available in Baveye and Vermeylen (1993).

In this initial list of 1280 students, but at the exclusion of the Cornell graduate students, three hundred names (i.e., 23.4%) were selected randomly, using a random number generator, under the constraint that no two students in the short list would have the same graduate advisor. In practice, when a violation of this requirement occurred, the name of the second student chosen was discarded and the random selection process was repeated one more time. The random nature of the sampling as well as the relatively large size of the sample, compared with the total population, insured the absence

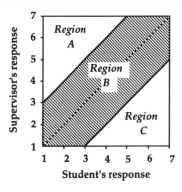

Fig. 10-3. Schematic illustration of the three regions used in the analysis of differences in perception between advisees and advisors.

of significant bias. Indeed, the composition of the sample of 300 students closely approximated that of the total population (comparison in Baveye & Vermeylen, 1993).

Method of Analysis

In order to identify possible divergences of perception between faculty supervisor and student, a graph like that of Fig. 10-3 was plotted for each one of the first 33 statements of the survey instruments. These graphs were then divided into three regions. A point falling in Region B means that the response of the advisor and of the student differed by at most two units on the Likert scale, suggesting a reasonable concordance of views. In Regions A and C, the differences are strictly larger than two and indicate a nonnegligible divergence of perception.

Which one of Region A or C was retained in the final analysis of the data varied from statement to statement. For most of the positively-worded statements, the points in Region A were retained because they were the only ones that were considered to have the potential to lead to conflicts. On the other hand, for these same statements, the points in Region C have probably little negative impact on the advisor–advisee relationship. This situation was of course reversed in the cases of the negatively-worded statements (no. 7 and 24) or of the statements inviting low responses on the Likert scale (no. 20 and 22). There, Region C was retained instead. In two cases (no. 8 and 14), both Regions A and C were retained because both were considered to have the potential to lead to misunderstandings between advisor and student. For example, for statement no. 14, severe miscommunication would occur if the advisor considers that he or she pays a lot of attention to the student as a person, while the student feels that his or her advisor could not care less. On the other hand, there is a risk of perceived intrusion if the advisor considers that he or she pays little attention to the student's life, while the student feels exactly the opposite.

For each statement in the first part of the survey instruments, the final result of the above analysis was expressed as the percentage of critical points (in Region A or Region C, or in Regions A and C) compared with the total number of data points for that statement.

Other statistical analyses, like the Pearson correlation analysis, were also carried out on the data. Further details on the methods used and on the results obtained are provided in Baveye and Vermeylen (1993).

RESULTS AND DISCUSSION

The average rate of return of the students' survey instruments was 55%, while that for the faculty supervisors was significantly higher, at 72%. Various reasons for this difference are analyzed by Baveye and Vermeylen (1993). They include the much higher mobility of the students as well as the fact that some departments do not seem to update their students lists on a regular basis. In 139 cases, survey instruments were completed and returned by both the student and his or her advisor. Elimination of pairs where, for various reasons (e.g., low proficiency in English, minimal interaction between official advisor and student), the responses were not considered very reliable, reduced the sample number to 128. The composition of the student body in this pool was reasonably similar to that of the total population (Table 10-1). Except for Group I, all the female groups were more represented in the sample than they were in the general population. This translates into a gender bias of 5.3% in favor of the female students. In comparison, the nationality and degree biases were very small; the first was only 0.6% in favor of the foreign students, while the second amounted to 1.5%, in favor of the M.S. students.

On a number of the forms, comments had been written in the margin, indicating that some of the statements had been considered ambiguous or unclear. Whenever this was the case, the responses to the particular statement in question were not entered in our database, to avoid any confusion.

With respect to the major functions of the faculty supervisor, as defined by Brown (1968) (Fig. 10-2), there is a remarkable similarity of views among advisors and students. In both populations, slightly more than one-half of the respondents chose Function 2 (Fig. 10-2) as the most important

Table 10-1. Composition of the sample and of the total population of soil science graduate students. (Nationals are defined as Canadians in Canada and U.S. citizens in U.S. institutions).

Group	Description	Soil science students population	Sample
		%	
I	Female, national, M.S.	11.3	7.0
II	Female, foreign, M.S.	3.7	8.6
III	Male, national, M.S.	24.6	25.8
IV	Male, foreign, M.S.	8.7	8.6
V	Female, national, Ph.D.	5.6	9.4
VI	Female, foreign, Ph.D.	4.6	5.5
VII	Male, national, Ph.D.	19.3	18.0
VIII	Male, foreign, Ph.D.	22.1	17.2

function, with ≈ 40% choosing Function 3 instead. The responses of the advisors and of the students were not significantly different.

The results of the analysis by regions have already been provided earlier. In the list of 33 statements in the previous section, we have written next to each statement the percentage of responses that suggested a strong divergence of perception between advisee and advisor.

Since the statements were classified according to 10 general factors, one way to analyze these percentage data would be to calculate their mean for each factor and to rank these means. In this case, career preparation would come ahead (with a mean of 26.4% of disagreements), followed by interpersonal rapport (23%) and respect for students (17%). Surprisingly, supervisor accessibility is at the bottom of the pack, with only 9.5% of disagreements. As informative as this quick analysis may be, it is, however, limited by the fact that, as we have already mentioned earlier, the factors used above to classify the 33 statements are to a large extent arbitrary.

A better approach consists of analyzing individually the statements with the highest percentages of disagreements. The five statements with the highest percentages of apparent disagreements are:

8. The advisor decides in detail what is to be done in the research and how it is to be done. 30.9%
32. The advisor is concerned about preparing the student for all aspects of the student's future career, not just for the research. 29.8%
14. The advisor pays attention to other aspects of the student's life besides the student's studies. 28.7%
13. The advisor helps the student to make contacts that could be useful for the student's career. 27.3%
22. In publications and talks, the advisor takes personal credit for the student's work. 22.9%

As with any survey instrument, part of the disagreements may stem from the way the statements are worded. One should therefore interpret these results with caution. Nevertheless, the disagreements are in some cases so severe, with a 7 mark for the advisor and a 1 for the student, or vice-versa, that it is hard to imagine that they could be ascribed entirely to a problem of semantics. While the actual numbers should probably be taken with a grain of salt, the above percentages suggest that the aspects of the advisor–advisee relationship addressed by these five statements should be the object of a special effort of communication.

In the five statements above, it is interesting to note that two (no. 32 and 13) deal specifically with preparing the students for their future career. To some extent this may be a reflection of the current economic crisis; the bleak job market makes students somewhat nervous about their future employment, after they leave the university. They feel apparently that their advisors are not doing sufficiently to address their concerns in this respect.

Almost 23% of the studies surveyed disagreed, sometimes strongly, with their advisor concerning the level of personal credit the latter takes for the student's work (statement no. 22). This difficult question of ownership of

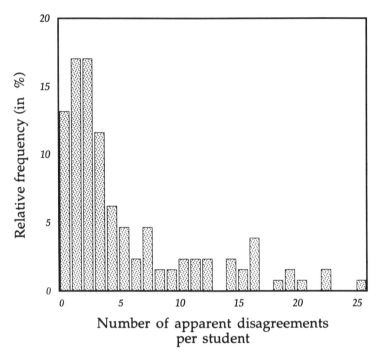

Fig. 10-4. Distribution of the number of apparent disagreements per student, based on the students' and advisors' responses to the 33 paired statements of the survey instrument.

the results of the graduate research has been addressed in the past by a number of authors (e.g., Bargar & Duncan, 1982). They suggest that in practice ownership of the research often is shared. It is therefore as unrealistic for the student as it is for his or her advisor to claim all the credit, as tempting as it may be for either of them to do so. The above result suggests that in close to 25% of the cases, advisees and advisors have not reached any kind of consensus on their respective level of ownership of the research outputs.

A possible criticism of the above percentages is that they are perhaps the result of a fraction of the students having a very conflictual relationship with their advisors, while the vast majority of the students have no communication problems whatsoever. If we plot the number of apparent disagreements per student (Fig. 10-4), it appears however that the students having a large number of apparent disagreements with their advisor are a small minority. At the other end of the spectrum, the students with absolutely no apparent disagreements with their advisors are also in a minority (at 13%). The majority of the students have a small (<5) number of disagreements. The latter, in first analysis, do not seem correlated with gender, nationality or degree program (further details in Baveye & Vermeylen, 1993).

The above analysis suggests clearly a number of areas where there is room for improvement in the communication between advisees and advisors. On the basis of these results, one would expect that, in the satisfaction

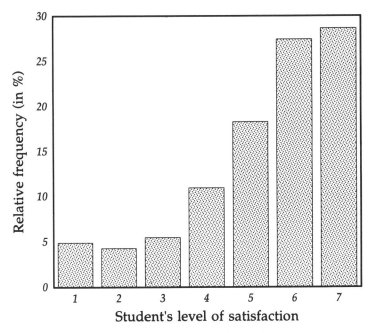

Fig. 10-5. Distribution of students' responses to question no. 34: How satisfactory do you find faculty supervision?. The level of satisfaction ranges from very unsatisfactory (1) to very satisfactory (7). All available students' responses were used to establish this diagram.

part of our survey, the students would appear dissatisfied or at least somewhat dissatisfied with the relationship they have with their advisor, and vice-versa. In fact, it is surprising to find that the opposite is much closer to the truth. In a vast majority of the cases (Fig. 10-5), students declared themselves satisfied and, for 30% of them, very satisfied with their advisor's supervision. At the same time, the advisors were in general very pleased by their relationship with their students and with the latter's performance. For example, in response to the statement no. 37 (Fig. 10-2), 42% of the faculty supervisors found very satisfactory the level of independence of their students (Fig. 10-6). Similar results were obtained for all other statements in the second part of the survey instruments. These observations, however, have to be taken with caution at this stage. Further analyses (reported in Baveye & Vermeylen, 1993) are needed to assess whether this euphoric outlook is real or only apparent.

ACKNOWLEDGMENTS

Sincere gratitude is expressed to Tom Bruulsema and Kirsten Verburg (Cornell University) for their help in pretesting the survey instrument. The comments and constructive criticisms of Gaye Burpee (Michigan State Univesity), Shin Watanabe (Cornell University), Drs. Larry Boersma (Oregon

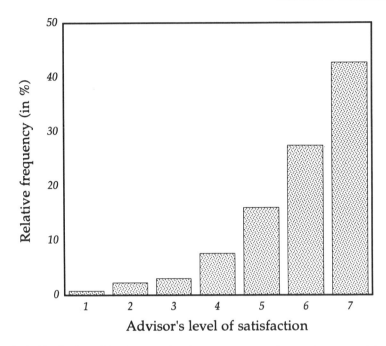

Fig. 10-6. Distribution of faculty supervisors' responses to question no. 37: How satisfactory do you find the level of independence of the student?. The level of satisfaction ranges from very unsatisfactory (1) to very satisfactory (7).

State University), and David B. Mengel (Purdue University) were very helpful in the analysis of the survey data.

REFERENCES

Barger, R.R., and J.K. Duncan. 1982. Cultivating creative endeavor in doctoral research. J. Higher Educ. 53:1-31.

Baveye, P., adn F. Vermeylen. 1993. Student body profile and advisor-advisee relationship in graduate soil science education in the U.S. and Canada: Survey results and analysis. SCAS Research Series No. 93-001. Cornell Univ., Ithaca, New York.

Best, J.W. 1970. Research in education. Prentice-Hall, Englewood Cliffs, NJ.

Brown, B.F. 1968. Education by appointment. Parker, New York.

Rugg, E.A., and R.C. Norris. 1975. Student ratings of individualized faculty supervision: Description and evaluation. Am. Educ. Res. J. 12(1):41-53.

Sorenson, G., and D. Kagan. 1967. Conflicts between doctoral candidates and their sponsors, a contrast in expectations. J. Higher Educ. 38(1):17-24.

Wiersma, W. 1969. Research methods in education. J.B. Lippincott Ct., Philadelphia.

11 Educational Needs in Soils and Crops of Graduate Students from Developing Countries[1]

W. E. Larson, R. Kent Crookston, and H. H. Cheng

University of Minnesota
St. Paul, Minnesota

ABSTRACT

A questionnaire was mailed to 216 current and former graduate students who had received or were pursuing degrees in soil and crop sciences from five U.S. universities. The objective of the questionnaire was to obtain an expression of opinion on pertinent issues related to graduate and continuing education from current and former students from developing countries. The questionnaire was divided into four sections, namely, purpose and support, background, graduate program, and follow-up. Responses were received from 127 people from 49 countries. Most of the respondents were satisfied with their classes and thesis research training. From the questions and the narrative comments it was apparent that they would have profited from a broader experience in the USA. Better exposure to the Land Grant University system of teaching, research, and extension is needed. Few had training or experience in communicating with farmers and almost one-half said they had no exposure to practical U.S. agriculture. Some thought that training or additional experience in classroom teaching would have been desirable. A number thought that they would have profited from working with their advisor and in the depàrtment in a leadership role so that they could apply that experience in their home country. After returning to their home country they felt closer contact with U.S. universities would be extremely helpful. Sabbaticals, visits to the USA, exchange visits to their country by U.S. scientists, attendance at international meetings, newsletters, and exchange of reprints were suggested. Clearly, U.S. university faculty need to be alert for opportunities to broaden student experience while in the USA and to continue contact with the degree recipients after they have returned home.

Meeting the needs of a quality graduate education for agronomic students from both developing and developed nations has long been a challenge for U.S. universities. While training procedures in basic science, research and

---

[1] Contribution from the Minnesota Agric. Exp. Stn., Paper no. 20170 Sci. Journal Series.

Copyright © 1994 Soil Science Society of America, 677 S. Segoe Rd., Madison, WI 53711, USA. *Soil Science Education: Philosophy and Perspectives.* SSSA Special Publication no. 37.

teaching are fundamentally the same for all students, the backgrounds of students from the developing countries, and future application of their training are often quite different when compared with students from developed countries. In agronomy, a natural resource profession, agroecological differences among regions of the world, as well as social, cultural, and political differences make application of information to solve human and environmental concerns extremely complex.

Many questions have arisen over the years as to how graduate programs in agronomy should be structured to meet the needs of international students. One group of concerns includes the degree to which training in basic vs. applied science, use of expensive equipment vs. practical equipment, field vs. laboratory research, and research vs. teaching should be emphasized. In deciding what emphasis should be given to an individual student or to a graduate program, one should consider the needs of the country or geographical region in which the scientist is going to work.

A number of studies have addressed the educational needs of students from developing countries, both at the graduate and undergraduate levels (Williams et al., 1963; Lee et al., 1981; Arnon, 1989a, b; Caddel, 1991). While educators' experiences and thoughts have most often been reported, some have also included results of student surveys (Lee et al., 1981); however, none of the surveys to our knowledge, have been solely concerned with the need of graduate education for students in soils and crops.

QUESTIONNAIRE

To help us in analyzing the educational needs of graduate students from developing countries, we designed a questionnaire for current and former students from developing countries to obtain an expression of their opinion on pertinent issues related to graduate and continuing education. The issues included in the questionnaire were based on the authors' experience with international students and graduate programs. Because of time limitations between the invitation and manuscript due dates, the questionnaire was purposely brief to ensure a good return rate. The questionnaire was sent to 216 individuals from 49 countries, all of whom were students or former students in soil and crop sciences at the Univesity of Minnesota, Cornell University, North Carolina State University, Texas A&M University, and the University of Hawaii.

The questionnaire was divided into four sections: A. purpose and support (2 questions); B. background (6 questions); C. graduate program (15 questions); and D. follow-up (5 questions). In addition, a request was made for narrative opinions and suggestions as to how graduate programs could be improved.

Responses to the questionnaire were received from 127 individuals (59%). All questions were completed by most respondents, and 57% offered written suggestions. Table 11-1 summarizes the number of questionnaire respondents from different areas of the world; there was a good global distribution.

EDUCATIONAL NEEDS OF STUDENTS FROM DEVELOPING COUNTRIES

Table 11-1. Number of questionnaire respondents from areas of the world.†

Geographic area	Currently in graduate degree program	Have received a graduate degree	Total
Africa	7	20	27
Asia	8	13	21
Pacific, Australia New Zealand	0	4	4
Europe	9	4	13
Central and South America	14	28	42
North America (Canada, Mexico)	5	6	11
Southeast Asia	1	8	9
Total	44	83	127

† If the student had received a M.S. degree from a U.S. institution and was now working on a Ph.D., the student was counted as being a current student.

The responses to the questionnaire were not compared by areas or degree status because the numbers did not represent an adequate statistical sample. A cursory examination of the responses, however, suggested no major differences between regions or degree status.

Table 11-2 presents a profile of the respondents broken down into crop and soil scientists and current and former students. A total of 53.4% were soil scientists and 46.5% were crop scientists. Thirty-two percent were current students and 68% were former students. Of the former students, 50.8% of the total's terminal degree were a Ph.D. and 17.0% were a M.S. degree. Most of the current students are working toward a Ph.D. degree.

Table 11-2. Profile of questionnaire respondents.

Classification		Percent
	Soils	
Current students working on:	M.S.	2.5
	Ph.D.	11.9
	Total	14.4
Past students receiving:	M.S.	6.8
	Ph.D.	32.2
	Total	39.0
	Crops	
Current students working on:	M.S.	3.4
	Ph.D.	14.4
	Total	17.8
Past students receiving:	M.S.	10.2
	Ph.D.	18.6
	Total	28.8

RESULTS AND DISCUSSION

The results of the questionnaire, our interpretation, and discussion follow in the order of the sections of the questionnaire. Each section first summarizes the results of the questionnaire and then presents the authors' interpretations and comments (interpretation). The questionnaire and a tabulation of the results are given in Appendix 1.

Purpose and Support

Results and Questionnaire

The purpose of Question 1 (Q1) was to test whether the student's primary interest initially was to be a problem solver in his or her home country or his or her aim was to become a scientist or teacher who could compete on the international market.

Sixty percent of the respondents considered that their pursuit for an advanced degree in a foreign country was to enhance their skills as a scientist or teacher; 33% to learn skills to better tackle agricultural problems in their home country.

Funding support for students (Q2) was predominantly from either the employer in their home country (26%) or the host institution in the USA (44%).

Interpretation

Because the responses were not exclusive of each other, we interpret the answers to Q1 to indicate that students want to be able to compete as an international scientist, while also helping to solve the problems of their home country.

We did not try to determine the sources of support provided by the host institution that funded their education; the original source of funds for many may have been a government agency such as USAID.

Background

Results of Questionnaire

The objective of this section (Q3-8) was to obtain an expression of the proportion of students who felt handicapped at the start of their graduate study because of inadequate command of the English language or scientific background.

Most of the respondents (88%) did not have English as a native language. Fifty-four percent had language training before starting graduate work in the USA. Most of the respondents met all English requirements (70%), and were not required to take additional English training after entering graduate school in the USA. Understanding English in classes or communicating with faculty was either a minor problem or not a problem from the outset

(90%); however, 10% indicated communicating in English was a major problem. A large majority of the students felt that they had an adequate background in basic sciences (91%) and in crop or soil sciences (81%). In the narrative responses, a number recommended a light classroom schedule be taken during the initial months so that more time would be available to improve their English and to become oriented and adjusted to the institutional and cultural environment.

Interpretation

Our experience is that communicating in English is less frequently a major handicap than it was a few decades ago, however, to a few students it is still a severe problem. Faculty should be alert to the need for additional formal training in English as well as the need for language practice with other students and faculty. Starting graduate work without an adequate knowledge of English can be traumatic. A few respondents felt that American professors should have more familiarity or appreciation of languages other than English.

It is sometimes difficult to determine accurately a student's adeptness in basic science because of difficulty in interpreting the curriculum from their home country. While our survey was not extensive enough to quantify the data by countries or sections of the world, our experience is that there are differences. In case of doubt, advisors should query or test the student's skills and suggest remedial classes, if necessary.

Graduate Program

This section consists of 15 questions. They can be grouped into the following interrelated categories: experience and preparation at the start of their graduate education (Q9-10), quality of their research experience (Q11-15), research location (Q16-18), other experiences (Q19-23). Since some of the questions may have implications to more than one aspect of graduate program, our commentaries have not always been limited to the categories under discussion.

Experience and Preparation

Results of Questionnaire. In Q9, we queried the respondents as to their experiences when they first started graduate school. Our hypothesis was that U.S. faculty could provide additional help when the students first arrive. About one-half of the respondents indicated they were not given sufficient orientation to U.S. graduate educational programs. Improved counseling by their advisors could have mitigated the problem greatly.

In the narrative, some respondents felt that U.S. university faculty could do more to help the international students get settled when they first arrive in the USA. Housing, shopping, local customs, language, schools for children, local transportation, and other important concerns were mentioned.

Faculty committees, student committees, peer mentoring groups, written material, and group orientation sessions could help.

The U.S. graduate programs generally have more emphasis on formal classroom training with less time for thesis research then European universities. In Q10, we queried the respondents as to their perception on the balance between classes and research. Although a majority of respondents felt that the balance between formal classes and thesis research in the U.S. university they attended was about right (69%), 23% thought that more emphasis should be placed on thesis research, and only 6% thought more emphasis should be placed on class work.

A common narrative suggestion was that classes, colloquia, and discussions on international and tropical agriculture should be incorporated into the U.S. curriculum. This suggestion offers benefit to both U.S. and international students, who learn from one another as they contribute to lectures and discussions.

Interpretation. Graduate students from developing countries are often required to make major adjustments when entering graduate school in the USA. The socioeconomic conditions, educational system, differences between undergraduate and graduate training as well as language often require major adjustments. Perhaps our emphasis on synthesis rather than rote learning is most different. Our experience has been that students often do not know what is expected of them when first entering graduate school.

We agree with the respondents that the balance between class and thesis time is about right, although the balance may vary with the student. Some foreign Ph.D. students come to the USA with considerable research experience. Their need may be greatest in formal classes.

Many U.S. professors need more familiarity with the culture and agriculture of developing nations. The ecosystems of the nations from which the students came were often tropical with both dry and humid climates. Landscapes were often characterized by steep slopes. Crops from these regions are different with a wider variety of species requiring different nutritional and cultural practices than in the USA. Field units are sometimes small and are tilled with implements that are powered by small tractors, animals, or humans. Farming practices, such as intercropping, fertilization, and residue management are often very different. First hand experience with these conditions would aid materially in selection of more appropriate thesis research projects and examples for classroom discussion. An appreciation of the culture of the developing nation by Americans makes the student feel more at ease and offers a rich background for discussion. As they should be, students are proud of their native culture. We are supportive of this suggestion made in the narrative comments that each international student be given an opportunity to present a seminar on his or her home country during their first year in the USA.

After returning home, many students from developing countries were placed in responsible positions and were required to develop broad research or teaching programs. Caddel (1991) suggests that the U.S. land grant system of teaching-research-extension partnership works well, but is often not understood by international students. A number of respondents felt that students should be given more opportunities to participate in the planning and conducting of their advisor's or department's programs so that they could carry out these responsibilities in their home countries. This can be an important learning experience that is often overlooked. How to plan a multiyear program, how to secure and allocate resources, and how to prepare research grant proposals and user reports are a few of the items that many students are faced with when starting a career position.

Research Experience

Results of Questionnaire. The purpose of Q11 through Q14 was to determine the respondents' feelings about the quality of thesis research counseling and to explore their thoughts on selection of a research topic. Most students felt counseling by their advisor in choosing classes, thesis topic, conducting research, and in writing a thesis was adequate; however, $\approx 36\%$ felt counseling in preparing for preliminary exams was insufficient (Q11). Students seemed to feel that the approach to their research needed to be both basic and applied, and that both are important to their education as well as to the welfare of their country (Q13-14). Students did feel technological differences (equipment, instrumentation, and facilities) should be considered (Q15). In choosing a graduate school, students were somewhat divided as to whether agroecological differences between regions of the USA and their home country were important (Q16).

Interpretation. Occasionally, U.S. universities are criticized for having foreign students work on basic research requiring the use of sophisticated equipment, when the need in their home countries is for applied research that can have a more immediate impact. While individual cases vary greatly, students from developing countries should obtain experience in solving current problems, and their thesis research should attack applied problems in a fundamental way. Basic understanding of research necessary to solve an applied problem should be emphasized, but the research should be carried through to fulfill a practical need.

A few students commented in the narrative that students should select universities located in similar agroecological situations. On the other hand, the perceived quality of the university in the student's area of interest and the reputation of the advisor were often the overriding consideration. If special curricula are wanted, obviously a school with that specialty should be sought.

Research Location

Results of Questionnaire. We asked the respondents in Q16 through Q18 about the importance of being at the same location as their advisor during the (i) planning, (ii) data gathering, (iii) interpretation, and (iv) writing stages (Table 11-3). Students felt strongly that they must have close contact with

Table 11-3. Importance of the student being at the same location as their advisor during four stages of theses research.

Stage	Very important	Moderately important	Not very important
		% of respondents	
Planning	91	9	1
Data gathering	23	48	30
Interpretation	71	27	2
Writing	57	40	3

their advisor during the planning stage (91%, very important). During the data gathering state, 23% said very important, and 48% said moderately important. Seventy-one percent indicated it was very important to be with their advisor during the interpretation stage and 97% felt it very important or moderately important to be at the same location during the writing stage. Clearly the majority of the students felt it best to be at the same location as their advisor during all stages of their thesis research.

Interpretation. Considerable debate has occurred over whether students from deveiloping countries should spend all of their time during graduate study in the USA or whether they could do their class work in the USA, take their examinations, and then return home to do their research and write the thesis. Proponents of the latter plan argue that it is less costly and insures that the research is relevant to the home country's problems.

Our experience is that thesis research goes much more smoothly if the student and advisor are together in the USA during all stages of the research. Depending on the individual case, however, excellent thesis research can be done if the student is in their home country during the data gathering stage. Several conditions are desirable, however. First, the necessary facilities must be available. Second, the advisor should spend time in the home country of the student at the time of research initiation plus one or two follow-up visits. We strongly recommend that international students return to the USA for data interpretation and writing. The benefits are obvious. One problem that has occurred in cases where the students return home for the data gathering stage is that they are often overloaded with demands to fulfill the duties of their regular positions, so that little time and effort could be devoted to their thesis research.

Other Experiences

Results of Questionnaire. In Q19 through Q21 we asked the respondents about the need for teaching and extension experiences in addition to research while in the USA and in Q22 through Q23, we asked about the importance of their graduation education and how well prepared they were for positions after completing their degree.

Most students received some classroom teaching experience, about one-half received training or experience in practical U.S. agriculture, but few received training or experience in activities that improved their skills in extending research information to farmers and others needing the information (Q20-21).

An overwhelming majority of the graduates felt they were well-prepared for their first or present position (Q22-23). They were unanimous in feeling they were well-qualified to do research (100%) but some indicated they were not so well-prepared to teach in a university (15%) and participate in other professional activities (20%).

Interpretation. The Ph.D. and even the M.S. degrees in U.S. universities are primarily research degrees; however, agronomists are often expected to do classroom teaching and have contact with users of their information (extension). Some narrative comments suggested that students should receive more training and actual teaching experiences. Experiences in how to structure a curriculum and a specific class, testing, and discussion activities were suggested. Several indicated that students would profit from more on-farm experience, and exposure to practical U.S. agriculture. A week or more of living on a U.S. farm experiencing the day-to-day operations should be explored for some students.

Follow-up

Results of Questionnaire

The purpose of this section was to explore the respondents' views concerning their continuing education after completing their degree requirements in the USA. Clearly, the respondents in the survey wanted more contact with their U.S. colleagues after they received a degree. Seventy-four percent felt a post-doctorate position would have been helpful, although only 11% had post-doctorate experience. A sabbatical leave in the USA within 5 to 10 yr following receipt of their degree was favored by 93%. All of the respondents said that a sabbatical would be very or moderately helpful. Ninety-one percent felt continued collaborative research with their advisor or other U.S. faculty was desirable.

Interpretation

Perhaps one of the weaknesses in U.S. university education for students from developing countries has been inadequate follow-up once they return home. Development of a scientist is a life-long process requiring continued learning. Post-doctorates, sabbaticals, visits by their U.S. professors, frequent visits to the USA by former students, and collaborative research projects, exchange of reports, or newsletters, should all be explored. Probably the most limiting cause for inadequate follow-up is that granting agencies have not allocated sufficient funds for these activities. We hypothesized that research or education programs could be strengthened measurably if more continued contact between U.S. and developing country scientists could be achieved.

SUMMARY AND CONCLUSIONS

The results of the questionnaire, including the narrative comments, indicated the following:

1. International scientists who received an advanced degree in soils or crops from a U.S. university were satisfied with their classes and thesis research.
2. They would have benefited from additional counseling concerning expectations of faculty, cultural adjustments, and personal matters.
3. They should be given a broader experience in the USA. Better exposure to the land grant university concept in teaching, research, and extension would be useful.
4. They expressed a strong sentiment that exposure to leadership in research, teaching, and extension would be helpful. A number felt that they would have profited from working with their advisor and in the department in a leadership role so that they could apply that experience in their home country.
5. Classroom teaching training and experiences were mentioned as an additional need in graduate education.
6. A strong appeal was made for closer contact between their U.S. faculty and the degree recipients after they return home.
7. U.S. professors need more familiarity with the agriculture, culture, and languages of developing nations.

Graduate advisors and departments should be alert for opportunities to give students broader experiences in all facets of agriculture. The extent and choice of the experiences should be worked out between the student and advisor. It may take additional time by the advisor and, in some cases, may detract from the students research time, but in our opinion, the rewards will be many.

APPENDIX 1

A questionnaire regarding the educational experiences and needs of graduate students from developing countries in the USA was developed by the authors with help from a faculty member from the Center for Survey Analysis, University of Minnesota. The questionnaire was sent to international students and former students in soils and crops from the University of Minnesota, Cornell University, North Carolina State University, Texas A&M University, and University of Hawaii. Of the 216 to whom the questionnaire was sent, 59% responded. Of the respondents, about one-third were current students, and two-thirds were past degree recipients, mostly Ph.D.s. Because of the availability of mailing lists, about one-half were students or former students from the University of Minnesota. A very few received their degree >10 yr ago. The respondents are from 49 countries.

The questionnaire follows. The numbers report the percentage of positive responses. More than one-half gave narrative responses to question 28, which are not included.

Graduate Education for Students from Developing Countries

A. Purpose and Support

Q1. What was your purpose in pursuing an advanced degree in a foreign country?
- 33.3 to learn skills to better tackle agricultural problems in your home country;
- 60.3 to enhance your potential as a scientist or teacher;
- 1.6 to advance in rank in your institution;
- 4.7 other (please describe).

Q2. Who funded your education in the USA?
- 26.0 my employer in my home country (institution or government);
- 3.2 personal or family finances;
- 42.5 host institution in the USA;
- 20.4 an international agency;
- 7.9 other (specify).

B. Background

Q3. Before starting your graduate work in the USA, what special training did you receive?
- 54.3 language training
- 45.6 other (specify)

Q4. Was English your native language?
- 11.8 yes
- 88.2 no

Q5. Was understanding English a problem in your classes and in communicating with faculty?
- 10.2 a major problem
- 38.6 a minor problem
- 51.2 not a problem.

Q6. Did you meet all English requirements when you first arrived or were you required to take additional training?
- 70.0 met all English requirements
- 29.9 required to take additional training

Q7. Do you feel your background in basic sciences (math, chemistry, and physics) was sufficient when starting graduate work in the USA?
- 91.3 yes
- 8.7 no

Q8. Was your background in crop or soil science sufficient when starting graduate work in the USA?
- 82.4 yes
- 17.6 no

C. Graduate Program

Q9. The U.S. system of graduate education is somewhat different from education in many other countries. When you started your graduate work in the USA were you given sufficient orientation about the U.S. system?
- 51.2 yes
- 48.8 no

Q10. Please evaluate the balance between formal class work and thesis research in your American graduate program.
  6.3 should place more emphasis on class work
  24.6 should place more emphasis on thesis research
  69.0 current balance between class work and thesis research is about right

Q11. Did counseling by your advisor and committee provide sufficient help in the following areas?

	yes	no
choosing your courses	86.6	13.4
preparing for your preliminary exams	64.3	35.7
choosing your thesis topic	88.8	11.2
conducting research	90.2	9.8
writing your thesis	87.4	12.6

Q12. Graduate thesis research can be described as basic or applied. Assume that basic research would develop new knowledge and understanding of principles and processes, while applied research would be to apply knowledge to analyze and solve problems in your country
What was your goal in your thesis research?

	very important	moderately important	not very important
basic research	46.8	45.2	7.9
applied research	60.3	34.9	4.8

Q13. How important are basic and applied research in your country?

	very important	moderately important	not very important
basic research	28.8	53.6	17.6
applied research	90.4	9.6	0

Q14. Is it important to consider technological differences (i.e., differences in equipment, instrumentation and facilities) between the host country and your home country when selecting your research topic? How important is it to consider these differences?
  41.4 very important
  37.5 moderately important
  21.2 not very important

Q15. How important is it to consider agroecological differences (e.g., differences in soils, climate, or crops) between your home country and the sections of the USA (east, west, or north central) when choosing a university for graduate education?
  27.6 very important
  39.4 moderately important
  33.1 not very important

Q16. Thesis research may be done in the USA or in your home country. Rate the benefits for you doing your research in the USA and in your home country.

	very beneficial	moderately beneficial	not very beneficial
USA	54.5	40.8	4.8
home country	46.3	40.6	13.0

Q17. Graduate thesis research in the USA can be divided into the following stages: (a) planning, (b) data gathering, (c) interpretation of data, and (d) writing. Rate the importance of being at the same location as your advisor during each of these stages in your thesis research.

EDUCATIONAL NEEDS OF STUDENTS FROM DEVELOPING COUNTRIES

	very important	moderately important	not very important
planning	90.6	8.7	0.7
data gathering	22.6	47.6	29.7
interpretation	70.9	26.8	0.2
writing	56.7	40.2	3.1

Q18. How long do you feel should be spent going from the B.S. degree from your country to a Ph.D. in the USA?
    5.6 less than 3 years
    76.9 3 to 5 years
    17.5 more than 5 years

Q19. Did you receive training and experience in practical U.S. agriculture that is helpful in your home country?
    55.6 yes
    44.4 no

Q20. If training or experience in classroom teaching was one of your objectives, are you satisfied with your experience?
    50.4 yes
    11.2 no
    38.4 not applicable

Q21. Did you receive training or experience in communicating with farmers?
    12.7 yes
    87.3 no

Q22. How important is your graduate education to you as you consider the following aspects of your life?

	very important	moderately important	not very important
your professional life	89.8	10.2	0
your education in basic principles	62.2	33.0	4.7
your ability to analyze real problems	59.5	37.3	3.2
familiarity with the research/teaching of the U.S. system	54.4	36.0	9.6
language and cultural aspects	41.6	38.4	20.0

Q23. After completing of your graduate education, did or will you feel prepared?

	yes	no
for your first position	90.4	9.6
for your present position (if different from the first)	84.1	15.8
to teach in a university	85.2	14.8
to do research	100.0	0
for other professional activities	80.4	19.6

D. Follow-up

Q24. Do you feel a post doctorate for 1 or 2 yr following your Ph.D. would have been/will be helpful in your education?
    74.4 yes
    25.6 no

Q25. Did you have a post-doctoral position?
    12.8 yes
    87.2 no

Q26. Would it be helpful to have a sabbatical leave in the USA or another country in 5 to 10 yr following your graduate degree?
 92.8 yes
  7.1 no
 If your answer is yes, how helpful would it be?
 82.2 very helpful
 16.9 moderately helpful
  0.8 not very helpful.
Q27. Would it be desirable to continue collaborative research with your advisor or other faculty from the university where you received your degree?
 91.3 yes
  8.7 no
Q28. What else do you think U.S. graduate programs can do to help graduate students from other countries?
 57% gave narrative comments.

REFERENCES

Arnon, I. 1989a. Kinds of research. p. 316-319. *In* Agricultural research and technology transfer. Elsevier, New York.

Arnon, I. 1989b. Formation of the research worker. p. 443-480. *In* Agricultural research and technology transfer. Elsevier, New York.

Caddel, J.L. 1991. Improving the education offered international students. J. Agron. Educ. 20:71-73.

Lee, M.Y., M. Abd-Ella, and L.A. Burks. 1981. Needs of foreign students from developing nations at U.S. colleges and universities. Natl. Assoc. for Student Affairs, Washington, DC.

Williams, David B. 1963. The development of effective academic programs for foreign students: Curricular, work experience, and social aspects. p. 123-128. *In* A.H. Moseman (ed.) Agricultural sciences for the developing nations. Am. Assoc. for the Adv. of Sci. Publ. 76. Am. Assoc. Adv. Sci., Washington, DC.

12 Advising Students from Developing Countries

Elemer Bornemisza
Universidad de Costa Rica
San Jose, Costa Rica

ABSTRACT

Graduate training of foreign students from developing countries is and might continue to be an important help received from the USA. Adequate student selection, based on academic and personal characteristics, a rather wide curriculum of studies, more than average advisor time, and not too short a study period usually produce graduates who will contribute to the progress of agriculture in their home countries. Thesis projects related to home country problems are generally useful. Advisor contact with recent graduates can contribute to better initial activities in the home country.

Graduate training of foreign students from developing countries is and might continue to be an important task of U.S. graduate schools and is an important assistance received by these countries from the USA. There is a long tradition in this field; I have worked with a number of professionals, trained half a century ago, who have made a considerable impact in many countries in the developing world.

There are four main activities in which recently trained scientists traditionally use their skills. Probably the most important is in upgrading and expanding ongoing research. There is indeed a great need for a more sustainable, and at the same time, more productive agriculture. A second common activity is to teach at the national institutions and prepare professionals who have an up-to-date view of agriculture. A third opportunity is to cooperate with international companies managing commercial plantations in the tropics. The availability of well-trained local technicians has reduced their dependance on expatriate employees. The last but not least opportunity is in managing the above three areas. The use of recent graduates as administrators is only a partial loss because while they themselves will contribute little in terms of research, education, or extension, they will have a much better understanding of the help and support they have to give to their collaborators who have advanced training in research.

Copyright © 1994 Soil Science Society of America, 677 S. Segoe Rd., Madison, WI 53711, USA. *Soil Science Education: Philosophy and Perspectives.* SSSA Special Publication no. 37.

GENERAL CONCEPTS

The American Society of Agronomy had an interesting discussion on this topic some 25 yr ago, during the Annual Meeting in New Orleans in 1968. Addressing the training of agronomists from abroad, Appleby and Furtick (1969) suggested six points required for excellent training, which appear to be as valid now as when they were proposed. These concepts are:

1. Foreign agronomists need high quality but *tailor-made* programs, considering the needs of the country and the person. Flexibility, well managed, is not synonymous with low standards.
2. Actual responsibility for some aspects of field research programs can often be a useful and new experience that gives confidence to the student. Special projects within the graduate programs and actual cooperation with the professor's project can be ways of obtaining this experience.
3. Training in the areas of administrative skills, effective communication, and good work organization can complement the acquisition of technical knowledge and can be very useful. Student participation in the projects where they take advantage of their advisor's personal experience can be very useful here.
4. The personal example of the professor who advises the student and his professional attitude can also be a very valuable experience. This is true for all students, but in some developing countries the distance between professor and student is so large that the personal contact in the USA is a new and useful experience for foreign students.
5. The importance of getting research results to the user is a concept many students have to learn and see working. Work experience in collaboration with extension specialists could be promoted as *special projects* within the study plan. This would slightly extend the study period of result in a lesser total amount of specific information assimilated by the students, but probably would be more useful in the long run.
6. Continued contact of the professor with recent graduates can contribute to better initial activities in the home country. Coddel (1991) puts emphasis on this post-training contact, institutionalized by some German universities. This could substitute, at least in part, for the post-doctoral experience, widely used in some fields, particularly in the basic sciences in the USA.

As with all teaching-learning processes the formation of students from developing countries depends on the interaction of the three components suggested by Martini (1990), which are students, teachers, and administrators.

Graduate study in a foreign country is a difficult task and students should have the interest, motivation, ability, and health for such an undertaking. If the students are married, which is quite common, the ability, interest, and dedication of the other partner is essential. Many able foreign graduate students have failed in the past due to family problems. The support of a dedicated and, if possible, well-prepared wife or husband can help the student to concentrate on his or her studies in a supportive surrounding. A well-designed and managed orientation program for both can be helpful. In this respect, the student advisor's wife or husband can be of great help in getting the student's family settled, in establishing contacts between the student's family and the surrounding community. If the student's religion has an established center at the university community, it can also be of some assistance.

An adequate knowledge of English is needed. This problem is now quite well taken care of by the generally obligatory TOEFL (test of English as a foreign language) test. In exceptional cases for students from countries where adequate instruction in English is difficult to find, it can be recommended that students with otherwise proven abilities be admitted without the necessary TOEFL score. In this case either short extensive English courses, or courses should be included in the program in parallel with the professional formation. Both methods have been tested and, properly used, have given positive results. Generally, students who have demonstrated good learning abilities in their home countries, as indicated by good grades, do well abroad in the USA also. As grading systems vary, it is very important that the student's grades are properly interpreted, which is sometimes difficult either because a system gives too low or too high grades. The writer was in charge of a graduate system where we had an interpretation code within which we could standardize the grades given in the dozens of universities from which we frequently received students. The interpretation code was prepared by a statistician, needed occasional revision, but generally worked well.

The role of student colleagues from the USA and abroad can also be very helpful. In many universities, more experienced graduate students help the new ones to find their way around and get established in the departments. This can be of particularly significant help for new students.

With regards to teachers, the advisor is a very important person for the success of foreign students. If he or she has time, more than average interest, and some knowledge of the student's country, his or her role will be determining. Generally, foreign students will be more time consuming than nationals. Communication problems take time to overcome and the foreign student's lack of knowledge of the university system will sometimes make it necessary for the advisor to provide explanations for concepts already well known to U.S. students.

With regards to administration, the two main requirements are a scholarship that provides the basic necessities of the student and time to complete his or her work. Adequate time might be somewhat larger than the average, particularly if the student comes from one of the least developed universities. A somewhat more extended time might allow some additional experience for the student such as collaboration with research or extension activities or with teaching if his or her English is good enough. The sponsoring agencies, official or private, should consider that one additional year might make a lot of difference. As a basic principle, two well-trained scientists can do much more than three endowed with a hurried training.

It is believed that four aspects are essential for successful graduate studies. Proper student selection, discussed previously, is an essential component and can be considered the first requirement without which nothing can be accomplished.

As a second point, a proper curriculum has to be identified. This will require more than normal time from the advisor since the educational background of foreign students is very variable and will be an essential component in the selection of the curriculum. A third point will be the possible future need of the student. Generally a wide curriculum resulting in considerable adaptability is needed. This might require more time, but probably will result in covering the needs of the home country where the recent graduate will be on his own, without always having the possibility of consulting specialists of related fields.

The role of an advisor as an example of a researcher is very important, particularly since the foreign student, when he or she returns to his or her home country seldom has the opportunity to work closely with and to emulate more experienced scientists. Appleby and Furtick (1969) examined in detail this non formal aspect of student advising and its great importance. Some universities have foreign student advisors whose mission is to help the students with the administrative aspects of their studies, e.g., visa renewal. Like the faculty supervisor, these foreign student advisors can also provide valuable assistance.

Time can be mentioned as the fourth essential element of graduate training. This is more or less critical in accordance to the student's background and financing. In most cases, it is a questionable procedure to try to reduce the training, resulting in a student who concentrates in narrow academic goals that allow early graduation, forcing him or her to miss many opportunities to learn topics not closely related to his curriculum but nonetheless useful for his formation.

GRADUATE STUDIES IN THE USA OR ELSEWHERE?

Probably most objective evaluators agree with Coddel (1991) when he affirms that the university system in the USA is generally doing a good job of educating students from developing countries. Convincing evidence for this is provided by the numerous contributions to research, teaching and even administration made in their home countries by alumni of U.S. universities. Some of the reasons for the success of the U.S. system are based on its good organization, its high level of manpower, the acquaintance of many professors with some of the practical problems encountered in developing countries, and generally its interest in providing assistance to developing countries.

Observers would also probably agree with Coddel's (1991) suggestion that the system can be improved, particularly by giving students experience and information on important topics not normally considered, such as research administration, planning, proposal writing, and others.

There are basically two alternatives to training in the USA for students from developing countries. One is to study in Europe, where many of the world's oldest schools of agriculture have produced excellent agronomists. The students who will feel comfortable in the European surroundings are those who have a very large dose of maturity and self-motivation. For these students, the European style of graduate work, where a student is typically expected to work independently, can give outstanding results. Unfortunately, these programs tend to last longer than those in the USA and the cost of living in Europe is high, which can lead to funding problems. Nevertheless, it is a good idea, if possible, when building up a research team in a developing country, to have members with different forms of training.

A second alternative is to use the new graduate programs in other developing communities, like Brazil and others. Experience with students obtaining their M.S. degree under such conditions is quite positive if the best schools in the group are identified and used. For doctoral programs, however, large faculties and facilities are needed and are often not yet present in these countries.

THESIS TOPICS FOR FOREIGN STUDENTS

At the graduate level the preparation of a thesis is an essential component of the program. What kind of thesis is most useful for students from developing countries? To answer this, one has to consider the purpose of a thesis in a graduate program. It is necessary for two reasons: one is training in research and for this the thesis has to be done properly, answering more to the question of *how* than that of *what*. For this kind of thesis the usual procedure at U.S. universities is high quality work that is expected of everybody, and is properly supported and oriented. In the case of foreign students, the orientation and supervision might take more time than usual to help the student to overcome an often weak and almost always very theoretical undergraduate training. Often the student has to learn for the first time to use field and laboratory equipment needed for modern research, some of which he or she has never seen before.

The second question of *what* responds to the need of a student not only to learn how to do research but to select an adequate problem and solve it properly. This second component initiates a researcher in a direction, often followed later on. For this second purpose thesis projects related to home country problems, and if possible, carried out in the home country, can be very useful. The criteria for selecting useful problems are often acquired by working with highly qualified colleagues, an opportunity not existing in many developing countries. As a result, if the advisor explains, even informally, how decisions on problem selection are reached, he or she can contribute to many years of useful future work.

This kind of research, often more applied than one which is typical for U.S. universities, needs some understanding of local problems by the student's advisor. If possible, the advisor should have an opportunity to visit the experiments. The presence of a local co-advisor, whenever available, can also be very useful. The problem is that sometimes when the student arrives home, his assistantship is discontinued and to survive he or she has to resume his or her former job, or an important part of it, thus delaying or inhibiting the work on the thesis. This can result in delays or the discontinuation of the study, and it should therefore be avoided as much as possible.

The option of orienting agricultural research towards a more basic or applied approach is examined by Arnon (1989) who suggests that adequate agricultural research should be oriented to give solutions to actual problems of the country of region the student is originating from. A home country thesis can contribute to this approach and get the student started along a line of work where in the near future he might contribute to the welfare of his country.

If no home-country thesis is possible, the methodology used should be the most adaptable possible, to avoid the problems of readaptation of the students in their home countries, as analyzed by Arnon (1989). The student who is encouraged to work on a problem that does not require complicated equipment will eventually be better able to avoid problems related to lack of institutional framework, lack of financial support or even the possible hostility of colleagues feeling threatened by a more *fancy looking* research.

If the student learns to work on problems that yield tangible results within a reasonable time, there is an increased chance of obtaining the required support in a timely manner for each research activity, making possible the establishment of a successful program.

Establishing research projects between the advisor and his former student, if they can be accommodated within the possibilities of both, could be very helpful for a new investigator.

Evidently it would be very useful if during the formation of the new researcher, two additional new ideas could be made part of his thinking. The first is team work. The importance of this in tropical agriculture and the necessity to train people in this respect was already indicated many years ago by Bradfield (1969). The considerable success of the green revolution to feed a large part of the world resulted from the team effort of a number of research centers.

Grobman (1969) insists on the importance to learn the necessity of the continuity of research. This can be well illustrated in the USA, but it is necessary to demonstrate and explain it also in developing countries. There is very little long term research in these countries even though it is definitely needed, particularly to promote sustainable agriculture. Young scientists, who can initiate and bring to completion research projects lasting 20 yr or more, should be motivated for and supported in this kind of endeavor.

Another ability, that of elaborating research projects based on adequately identified problems, is a crucial necessity and is unfortunately quite difficult to learn. A properly conducted seminar on agricultural research could be a way to teach it.

Learning to adopt the appropriate technologies, and to adapt to existing conditions, are important necessities in developing countries. Professors who worked under these limitations might be able to and at least should attempt to pass on to the students some of their experience in this field.

The capacity to organize one's research, or that of a group, is a necessity of many recent graduates, who seldom have the necessary information and experience for it. The graduate seminar on organization and administration of research, previously suggested, might be a solution for this problem. Texts like that of Arnon (1989) can also contribute useful information. The capacity and desire to communicate research results to colleagues, administrators, and users is fundamental. Also, good communication keeps teams together, and this is needed for the complex research programs of today.

Motivation and the feeling of the need and urgency of results should complete the list of the competencies of a good research worker for developing countries and should be stimulated in their informal education.

THE ROLE OF COURSE WORK IN GRADUATE STUDIES

The purpose of course work is to transmit current information to the student by different teaching methods so that he or she is prepared to handle the conceptual content and methodology of his or her field of specialization, and knows how to complement it later. It should also prepare him or her to understand articles published in the field.

As many students will return to positions where they are the one who knows most, the possibility to obtain help from colleagues is remote. As a result, they have to be largely self-sufficient. To achieve this, a rather wide curriculum is helpful, even if this reduces the depth of the information. Many U.S. graduate departments believe in an ample course work as indicated by Black (1972) and this is quite useful. In addition, the students should be trained so they can learn on their own, i.e., obtain the needed information, by developing a habit of reading the literature and incorporating it in their work, and passing it on to their coworkers and students.

Since many of the recent graduates will be promoted to administrative positions, they will have much use of information on communication and planning. It is useful to incorporate some introduction to these topics in the student's curriculum.

Sometimes the course work has to begin with introductory courses that are prerequisites for more advanced courses. Some students often find it easier to acquire prerequisite knowledge by taking courses than, as in the case of many European universities, via independent study; however, these courses, designed for undergraduates, are very labor-intensive and should only be used if absolutely necessary. Quite often, reviewing the corresponding texts provides the update the students need.

A general research seminar, well oriented and considering in addition to U.S. agriculture that of developing countries, could be a very valuable addition for the training of students from abroad, and also from the USA.

PROBLEMS WITH SOPHISTICATED EQUIPMENT

For graduate work, publishable research is required and since in many colleges of agriculture in the USA, state-of-the-art equipment is available, it is often used by foreign students for their thesis research. The dependence on up-to-date instruments that often results is one of the criticisms against recent foreign graduates in many developing countries. Cooperative arrangements for the use of advanced equipments with the student's former school can alleviate this problem and should be encouraged. It is suggested, if at all possible, that the student should also receive information on how measurements and chemical determinations could have been performed without using modern equipment.

This is a problem that is quite easy to handle, if there is an interest on the part of the faculty advisor to prepare the student for the generally simple conditions under which he or she will work in his or her home country. If this is done, no problems will appear. Students should try to acquire experience in putting equipment together, in their maintenance and their eventual purchasing; however, with regard to equipment problems in general, one should never forget that no equipment is better than its user or the maintenance it receives. This principle should be learned from the use of the instruments necessary for the research.

In addition to equipment selection, information on the organization of field stations, laboratories and data processing equipment can contribute significantly to the formation of graduate students.

CONCLUSIONS

In the past, the U.S. university system has done a good job of training students from developing countries. Many reports such as that of Rohweder et al. (1972) document this fact.

When the conditions of an able student, a dedicated advisor and a supportive administration are all satisfied, the system can contribute now, and will quite likely continue to contribute in the future, to form useful agricultural scientists for the development of the many countries who need it.

It is believed that the U.S. method is probably the most efficient organization for training students from developing countries, particularly at the Ph.D. level.

It can be observed that usually when the results were not the expected ones, some essential components of the method did not work properly. Poor student selection, inadequate support, and an unusually different system, compared with that in place in the student's home country, are some of the most common causes for failure. The large number of successful graduates of the U.S. educational system, however, indicates that, even under suboptimal conditions, useful scientists, teachers, and agricultural administrators are formed, who contributed, and are still contributing, to the progress of agriculture in their home countries and who, like the author, are very grateful for the training they have received.

REFERENCES

Appleby, A.P., and W.R. Furtick. 1969. Meaningful experiences for agronomists from abroad who attend U.S. universities for professional training. p. 21-25. *In* J.R. Cowan and L.S. Robertson (ed.) International agronomy, training and education. ASA Spec. Publ. 15. ASA, CSSA, and SSSA, Madison, WI.

Arnon, I. 1989. Agricultural research and technology transfer. Elsevier Applied Science Publ., New York.

Black, C.A. 1972. A perspective of graduate education in soil science: The future. J. Agron. Educ. 1:2-6.

Bradfield, R. 1969. Training agronomists for increasing food production in the humid tropics. p. 45-63. *In* J.R. Cowan and L.S. Robertson (ed.) International agronomy, training and education. ASA Spec. Publ. 15. ASA, CSSA, and SSSA, Madison, WI.

Coddel, J.L. 1991. Improving the education offered international students. J. Agron. Educ. 20:71-73.

Grobman, A. 1969. Scientist equipped for international agronomy. p. 65-78. *In* J.R. Cowan and L.S. Robertson (ed.) International agronomy, training and education. ASA Spec. Publ. 15. ASA, CSSA, and SSSA, Madison, WI.

Martini, J.A. 1990. A personal experience in the teaching-learning process. J. Agron. Educ. 19:66-71.

Rohweder, D.A., W.R. Kussow, A.E. Ludwick, and P.N. Drolson. 1972. Research and graduate training as a basis for promoting rapid change in traditional agriculture. J. Agron. Educ. 1:33-36.

13 Nontraditional Students: Off-Campus M.S. Degree in Agronomy

W. L. Banwart and D. A. Miller
University of Illinois
Urbana, Illinois

ABSTRACT

A statewide off-campus M.S. degree program was initiated in the spring of 1986 to provide continuing professional education for the nontraditional adult student. Groups targeted included faculty and staff at community colleges and high schools, extension personnel, technical and sales representatives of agricultural industries, and farmers. University of Illinois faculty offer courses at three to four locations around the state at any one time utilizing community college, area extension, and continuing education facilities. Requirements for admission are same as on-campus M.S. students. Faculty advise the 30 to 40 students typically in this program and supervise special problems or thesis research. Limitations of the program are library access and adequate student numbers at individual locations to justify offering advanced courss. A recent evaluation concluded the program is of very high quality, successfully meets the needs of both students and faculty, and will continue to provide a unique opportunity for advanced adult education in Illinois.

Trends in higher education today suggest enhanced interest and emphasis on the nontraditional student, in particular the older student. The American Association of State Colleges and Universities and the National Association of State Universities and Land Grant Colleges predict that by the year 2000, one-half of higher education students will be over the age of 25, and 20% will be 35 yr of age or older (Ludwig & Latouf, 1986). This same study reported that in 1986 <20% of the nation's college students were between the ages of 18 to 22 years, attending college full time, and living on campus. Another survey at the University of Texas at Dallas found the average age of undergraduate students was 29 yr (Galerstein & Chandler, 1982). Adults are returning to college not only to earn undergraduate degrees, but also are returning in increasing numbers to attend graduate school on a part-time basis, while maintaining regular jobs. In 1986-1987, part-time graduate enrollments accounted for more than one-half of the total graduate level enrollments at U.S. colleges and universities (Donaldson, 1991).

Interest by the nontraditional student in enhanced education does not apply only to recent graduates. The University of Illinois recently completed a survey of 1976 graduates 15 yr after graduation (Univ. of Illinois, Office of Planning and Budgeting, 1992, unpublished data). This study showed 23% of the graduates returning questionnaires had received one or more additional degrees since 1976, and that 14% were attending school in 1991, the year of the study. Additionally, almost 79% of the 1976 graduates reported they had participated in noncredit continuing education programs. The growth of adult education has affected administration, faculty, personnel services, and many other facets of colleges and universities. Many campuses especially in recent years of decreased enrollments have attempted to interest adult students in credit and noncredit offerings. Especially for the working adults, however, location and release time from work can prohibit travel to campus for additional course work or the pursuit of an advanced degree. Off-campus courses have been used to fill this void by many institutions but these courses seldom lead to an advanced degree. In 1968, the Cooperative Extension Service and Community College personnel in the state of Illinois requested that the Agronomy Department offer off-campus courses to support a M.S. program. Since that time many graduate courses were offered off-campus but until 1986 no more than three units (12 semester hours) of off-campus courses could be credited toward the M.S. degree. In 1986 a program that recognized the need for M.S. degrees for nontraditional students not having ready access to the campus was instituted at three locations in Illinois to permit the completion of a M.S. degree from the University of Illinois without attending *any* classes on campus (Miller & Schrader, 1989). Following a successful start this program has been continued on a statewide basis. The purpose of this chapter is to describe this Off-Campus Masters of Science (OCMS) program and to report the results on the effectiveness of this program from a survey of graduates.

PROGRAM DESCRIPTION

Purpose

The purpose of the OCMS program is to offer the possibility of an advanced degree in agronomy for the nontraditional student who, because of career, location, or personal constraints cannot enroll at the University of Illinois main campus (Urbana–Champaign). These are students who may be seeking to improve their professional skills, prestige, leadership opportunities, employment advancement, salaries, or job mobility by completing an advanced degree. They have included professionals in all phases of the agricultural sector such as agricultural industry, education, technical and sales personnel, farmers, and state and federal agencies.

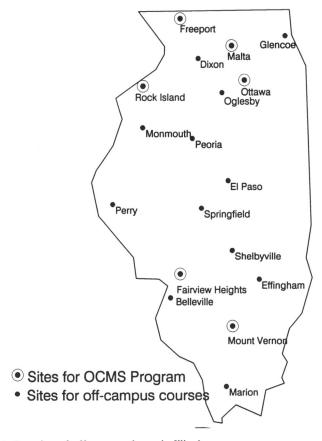

Fig. 13-1. Location of off-campus classes in Illinois.

Location

Geographic locations where the program is offered are selected in consultation with personnel from the extension service, community colleges, and continuing education regional offices of the University of Illinois. These sources are used to estimate the demand from potential students in a target geographical region in the state. Figure 13-1 shows the location of six sites where the OCMS program could be completed in 5 yr or less and 12 additional locations that could complement the OCMS program. The program is offered at only three to four locations at any one time. A main core of agronomy courses are offered that are supplemented by selected courses from the Departments of Entomology, Agricultural Economics, and Plant Pathology depending on the programmatic needs of a specific location. Courses taught at any one of these locations can be taken by nondegree students as well as those enrolled in the OCMS degree program.

Admission and Degree Requirements

Applicants are considered for admission if they have a baccalaureate or equivalent degree comparable to that granted by the University of Illinois, with a grade point average of at least a B for the last 60 h of undergraduate work and any graduate work completed. Students with limited training in agronomy or basic science courses may be required to take additional courses during their graduate programs. The admission decision is made by a faculty committee which, in addition to the grade point average, evaluates the quantity and quality of courses in the undergraduate program, three letters of reference, and a statement of interest prepared by the candidate. Both a thesis and a nonthesis option are available for M.S. candidates accepted to the program and minimum requirements for completion of the degree are identical to those for students studying on-campus. Minimum completion requirements for the two options include:

Thesis

1. Minimum of five graded units (20 h) of formal coursework approved by an advisory committee of the faculty.
2. Minimum of one graded unit (4 h) at the most advanced level (400 level) for the University of Illinois
3. No more than three units of thesis research.
4. Minimum of 1/4 unit (1 h) seminar.
5. Successful defense of a thesis.

Nonthesis

1. Minimum of eight graded units of formal coursework approved by a faculty advisory committee.
2. Minimum of three graded units of graduate study at the most advanced level (400 level).
3. Maximum of one unit of independent study under the supervision of a faculty member. This advanced study may consist of a field, laboratory, or other research problem consistent with the interests of the student, availability of facilities, and approval of the advisory committee.
4. Minimum of 1/4 unit (1 h) seminar.
5. Successful completion of a written or oral final exam.

Students in the OCMS program are assigned a faculty advisor when they are admitted to the program. The enrollment in the OCMS program has been ≈ 30 to 40 students at any given time. Faculty serve as academic advisors to these students and supervise special research problems or thesis projects. A toll-free line is provided by the Division of Extramural Courses that can be used by students to contact instructors, advisors, the Graduate College, Office of Admissions and Records, or other academic and administrative unit at the main campus. Individualized acadmic advising is also provided at off-campus locations by course instructors, and the OCMS coordinator as needed.

Instruction

Courses are most often taught as 3-h evening classes to accommodate the schedules of the nontraditional students, most of whom have full time jobs. Faculty travel via plane or automobile to the off-campus facilities that are coordinated by the Division of Extramural Courses. The state of Illinois has in excess of 30 community colleges positioned throughout the state that serve as potential sources of lecture and laboratory facilities. Other facilities include extension centers, high schools, and community buildings. Other delivery systems have also been used to take classes outstate including Telent (a conference type phone link) and an audiographic teleconferencing system that allows visual and audio linkage to multiple sites. Live personal interactive video is not currently available, but will make this type of degree program much more accessible and efficient in the future. Reserve materials are sent to libraries on location and students enrolled in off-campus courses are provided courtesy cards for the on campus library.

PROGRAM EFFECTIVENESS

In order to evaluate the OCMS program at the University of Illinois, a survey was sent to 47 students that were identified as currently enrolled or having recently completed degrees through the OCMS program. We received 39 replies to the survey with four responses indicating enrollment in off-campus courses, but *not* in the M.S. degree program (OCMS). *Only* results from individuals participating in the OCMS program are reported in this chapter. Following are summaries evaluating various components of the OCMS program at the University of Illinois.

Quality of Off-Campus Courses

Students in the OCMS program rated the quality of the OCMS program very commendatory (Table 13-1). Instructional quality was rated an average of 4.5 out of a possible 5.0 (4.5/5.0) and was the highest rating for the categories listed. Both course content and relevance of instruction to practical job skills received a rating of 4.3/5.0. This is encouraging considering the typical student is not only older but generally gainfully employed and therefore very cognizant of the relevance of instruction to their real world. The quality of instructional materials rated somewhat lower with approximately one-fourth of respondents choosing *average*. This may reflect some limitations of the variety of instructional materials an instructor may chose to use when such materials must be transported as much as 320 km (200 miles) or more to some locations; however, the university office of extramural programs makes every effort to accommodate any needs of instructors. Library

Table 13-1. Quality rating of off-campus courses by students.

	Very high quality (5)	High quality (4)	Average (3)	Low quality (2)	Very low quality (1)	Average rating
			% responding			
Instructional quality	49	49	2	0	0	4.5
Course content	34	60	6	0	0	4.3
Learning assignment/ experience	20	72	6	2	0	4.1
Relevant to practice	44	41	15	0	0	4.3
Instructional materials	18	56	26	0	0	3.9
Library resources	9	40	30	12	9	3.3
Overall quality	37	57	6	0	0	4.3

resources was by far the lowest category with an average rating of 3.5/5.0. Lack of extensive library resources has been identified in this and previous surveys as one of the limitations of off-campus instruction when compared with the on-campus experience. Courses are frequently taught at a community college where the library resources, especially in the agricultural areas, may be limited. All students in the off-campus courses are offered a card allowing access to the main library on campus as well as many other libraries throughout the state for the semester in which they are enrolled; however the distance of most students from the main campus prevents them from full utilization of this service. Students are also given a toll free phone number to access the main campus library where materials can be ordered by phone, but this survey clearly indicates students perceive the quality of library resources to be much lower than for the other services listed. Overall, students are very satisfied with the quality of off-campus courses with 94% of the respondents rating the quality of off-campus courses as either very high or high.

Students also believe that off-campus courses help them accomplish their individual learning objectives. Sixty-two percent responded very well and 38% well to the question "How well do you think the off-campus courses in which you enrolled accomplished your individual learning objectives?" (data not included). No students selected alternate responses, which were not very well or not at all.

Reasons for Enrollment

The survey also provided information about why these nontraditional students chose to enroll in off-campus courses and the OCMS program. The highest percentage of students chose as very important the opportunity to combine study with full-time employment (Table 13-2). Individual comments in this survey also emphasized the importance of evening classes in being able to complete the OCMS program. Also ranking high in reasons for enrolling in off-campus courses was the perceived quality of the program. This may have been the result of several factors including OCMS brochures and fliers, organized meetings with representatives of the university to explain

Table 13-2. Importance of selected factors in off-campus enrollment.

Factor	Very important (4)	Somewhat important (3)	Somewhat unimportant (2)	Very unimportant (1)	Does not apply	Average rating
			% responding			
Convenience of location	59	38	3	0	0	3.6
Convenience of schedule	38	62	0	0	0	3.4
Quality of program	73	24	3	0	0	3.7
Relevance to job	53	32	6	0	9	3.2
Opportunity to combine study with full-time employment	85	6	6	0	3	3.7
Relevance of courses to degree objective	47	50	3	0	0	3.3

the program, and word of mouth from friends and acquaintances previously involved in the program. Fifty-nine percent of respondents also identified convenience of location as very important. Students attending these classes are generally within 96 km (60 miles) of the location where classes are offered. Individual comments stress the fact that a M.S. degree would simply not be possible if classes were not brought to the student. More than 50% of the respondents also felt relevance to their job was a very important factor in their decision to enroll in off-campus courses. Slightly fewer students felt relevance of courses to degree objective was very important and only 38% of respondents chose convenience of schedule as being very important in the decision to enroll in off-campus courses. It should be noted that almost all classes in the OCMS program are offered in the evening or on Saturdays.

Importance of Support Materials or Services

Various supporting materials or services are offered for OCMS instruction. Student responses indicated instructional materials provided by the faculty are most helpful with an average rating of 3.6/4.0 (Table 13-3). A significant factor in support of off-campus instruction is the availability of an in-call toll-free line for students. This line can reach the campus extramural office where questions relative to admission status, requests for transcripts, and financial assistance can be answered. The toll-free line can also be used for academic advising sessions and student–instructor contact on course related materials. Students can, for example, seek help with problem sets or clarification of lecture materials from the instructor without the cost of long distance phone calls. The results of this survey (Table 13-3) indicate that while this service was deemed important (3.4/4.0 rating) some students (15%)

Table 13-3. Importance of various support services to off-campus instruction.

Service	Very helpful (4)	Helpful (3)	Not very helpful (2)	Not at all helpful (1)	Didn't know it was available	Does not apply	Average rating
			% responding				
Course planning	20	54	11	0	11	3	3.1
Student handbook	15	44	21	6	12	3	2.8
Personal counseling	35	47	6	3	6	13	3.3
On-campus visit	26	53	12	0	0	9	3.2
Toll-free telephone	44	24	9	3	15	6	3.4
Library resources	15	35	29	15	3	3	2.5
Instructional materials provided by faculty	65	35	0	0	0	0	3.6

were not aware of its existence. Students rated the library resources the least helpful of the services or materials evaluated. Limitations of library services for off-campus instruction were discussed earlier.

Benefits of the Off-Campus Courses

Students were also asked to evaluate their perceived benefit of off-campus courses in their professional development. The results (Table 13-4) indicate the greatest improvement in the range of techniques and skills students gain. Sixty percent of the respondents believe their skills improved a great deal by participation in this program. Students also reported increased job mobility (rated 3.3/4.0) and increased prestige among professional col-

Table 13-4. Effect of off-campus courses on selected factors.

Factors	Improved a great deal (4)	Improved somewhat (3)	Not improved	Decreased	Don't know	Does not apply	Average rating
			% responding				
Salary/income	3	40	34	6	17	0	2.5
Prestige among professional colleagues	17	66	3	0	9	6	3.2
Range of techniques and skills	60	34	6	0	0	0	3.5
Ability to influence your organization	0	66	14	0	11	9	2.8
Leadership within professional organizations	6	47	24	0	3	21	2.8
Job advancement where employed	17	23	17	0	6	37	3.0
Job mobility	21	41	3	0	12	24	3.3

Table 13-5. Evaluation of perceived career opportunities for graduates with a strong training in soil sciences.

	Excellent (1)	Good (2)	Fair (3)	Poor (4)	Very poor (5)
			% responding		
What do you believe are the career opportunities for graduates with a strong training in soil science?	10	54	32	3	0

leagues (3.2/4.0) as important benefits. The category receiving the lowest rating (2.5/4.0) was improvement in salary or income. Even so, 43% of the students reported at least some improvement in salary or income as a result of off-campus courses while 34% reported no improvement.

The survey also asked the question, "What do you believe are the career opportunities for graduates with a strong training in soil science?" Of the students responding to this only 10% indicate excellent career opportunities for persons with a strong training in soil science, while those expecting career opportunities for these students to be good or fair were 54 and 32%, respectively (Table 13-5).

An attempt was also made to determine which selected courses in the general area of soil science students believed were the most important in terms of training for their careers. Students responding included both those with soils as a primary interest ($\approx 20\%$) and those listing crop science as a major area of interest or whose primary interest was not listed. This ranking (Table 13-6) provides guidance to those wishing to offer soil science courses in an off-campus program. Numerical rankings of the top four courses suggest they were clearly more important to students than the others listed. The need for soils courses addressing environmental issues in soil science was deemed more important by students in this survey than courses such as soil chemistry, soil microbiology, or soil physics.

Table 13-6. Courses ranked in order of importance with respect to career training.

Course title	Ranking
Soil fertility	1
Basic soil science	2
Soil-plant relationships	3
Soil conservation	4
Environmental soil science issues	5
Soil testing	6
Soil chemsitry	7
Soil microbiology	8
Soil organic matter	9
Soil physics	10
Soil mineralogy	11
Soil physical chemistry	12
Research methods in soil analysis	13

Faculty Perspective

The data reported in this chapter has been from a survey of OCMS students conducted in the summer of 1992; however a comprehensive evaluation of the off-campus program including students in the OCMS program, students simply taking off-campus classes and faculty was conducted in 1989. Questionnaires were sent to 26 faculty involved in off-campus teaching and full or partial responses were received from 24 individuals. Results from this evaluation indicate that faculty also believe off-campus instruction is both useful and successful. Eighty-four percent of those responding believe the academic subject matter covered in off-campus courses is the same as that normally covered when the same course is taught on campus. Slightly less (74%) indicated the amount of material covered in the off-campus degree program was the same as that covered on campus. Some instructors indicated in written comments that material taught was adjusted to reflect the interests and academic background of students involved. Some also felt the off-campus students had a stronger agronomic background because of job experiences which allowed more material to be presented. Laboratory exercises were identified as very difficult to accomplish in most off-campus settings because of lack of facilities and difficulty in transporting materials by the instructor. Ten out of 18 instructors evaluating student academic performance rated off-campus students equal or superior to on-campus students, five instructors indicated quality was too variable to make a distinction, while only 3 of 18 believe the quality of the off-campus student is inferior to that of the students on campus. Approximately 90% of the instructors that responded believe the quality of their course taught off campus was equal or superior to the quality of the same course taught on campus. Thus in general, faculty in this survey found off-campus instruction to be satisfying and rewarding.

In summary, we have described an OCMS program at the University of Illinois that has been very successful. We believe it is a high quality program that has met the needs of students desiring a M.S. degree in agronomy but who cannot, for various reasons, return to the main campus. Perhaps the many expressions of the value and meaning of this program are related in the general comments of one student who wrote, "I found the program to be challenging and an opportunity to fulfill a life long dream that was not available to me otherwise. For the younger professional it is an opportunity to continue to earn a living, while advancing skills. In today's economy many of these people could not afford to take time-out to return to a campus setting. It also challenges the instructors. It is no easy task to cram a week's worth of instructions into a one night session. After a long day both the instructor and student are tired. The material must be geared to grab attention and keep it! The instructor must be enthusiastic and above all well versed in their field because the student's experiences will likely be brought into the discussion, unlike an undergraduate class with no work experience. My thanks to Dr. Miller (the OCMS coordinator) and the University of Illinois College of Agriculture for providing the vehicle for our minds to continue to grow whatever our age may be!''

REFERENCES

Donaldson, J.F. 1991. An examination of similarities and differences among adults' perceptions of instructional excellence in off-campus credit course programming. Innovative Higher Educ. 16(1):59-78.

Galerstain, C., and J.M. Chandler. 1982. Faculty attitudes towards adult students. Improv. College and Univ. Teach. 30(3):133-137.

Ludwig, M., and G. Latouf. 1986. Public, four-year colleges and universities: A healthy enrollment environment? Am. Assoc. of State Colleges and Univ. and Natl. Assoc. of State Univ. and Land-Grant Colleges, Washington, DC.

Miller, D.A., and L.E. Schrader. 1989. A statewide master of science degree program in Agronomy. J. Agron. Educ. 18:125-126.

14 Distance Education in Soil Science: Reaching the Nontraditional Student

Angelique L. E. Lansu, Wilfried P. M. F. Ivens, and Hans G. K. Hummel

Open University of the Netherlands
Heerlen, The Netherlands

ABSTRACT

The Open University of the Netherlands provides academic programs for open distance education. The courses that make up the programs are developed for adult students who are not able to attend courses at regular universities. A description is given of a soil science course offered by the Department of Natural Sciences. The course consists of printed material, allowing study to take place at home, and makes use of interactive learning materials. The central theme of the course is the variety of function of soils, within the perspective of environmental issues in the Netherlands. To enable the students to develop, in a distance-learning mode, the problem-solving skills needed in environmental soil research, an interactive video program forms an integral part of the course. It provides opportunities to the students to acquire and to analyze information from a number of sources, including soil-process models and Geographical Information Systems (GIS).

In December 1990, the Open University of the Netherlands (OU) began to develop a new soil science course. This course, to be launched in mid-1993, forms a part of the Environmental Science curriculum of the Department of Natural Sciences and differs in two main respects from regular soil science courses. First, its didactic approach makes the course suitable for distance education. Secondly, the focus of the course on environmental issues is in sharp contrast with the orientation on soil genesis and agricultural soils adopted in traditional courses.

This soil science course is described in detail in this chapter, which is organized as follows. First, a brief overview is given of the Dutch Open University, in particular of its educational format and of its targeted clientele. This overview is followed by a description of the philosophy behind the curriculum of the academic programs of the Department of Natural

---

Copyright © 1994 Soil Science Society of America, 677 S. Segoe Rd., Madison, WI 53711, USA. *Soil Science Education: Philosophy and Perspectives.* SSSA Special Publication no. 37.

Sciences. The consequences of this philosophy in terms of both program contents and course development are analyzed. The main part of this chapter considers the recent changes in the role of soil science in society and of their implications for soil science education at the OU. The objectives of environmental soil science education are discussed in detail as well as the possibilities that exist to achieve these objectives in a distance-learning mode, by adopting new didactic techniques and new means of communication.

Although this chapter refers mainly to the situation currently existing in the Netherlands, it is hoped that it will provide useful ideas for soil science education in other countries.

OPEN DISTANCE EDUCATION: A CHALLENGE

The Netherlands have a system of higher education that is relatively affordable for students and is geographically dense, with neighboring universities rarely > 50 km apart. Nevertheless, enrollment of members of the lower socioeconomic classes, of women and of disabled persons tends to be traditionally low in higher education.

The OU was founded in 1984 to address, in particular, the educational needs of these underrepresented groups. One of its goals is to make higher education accessible to adults who currently do not have, or did not have when they were younger, the opportunity to benefit from the programs offered by regular institutions of higher education. In so doing, the OU fulfills one of the ultimate objectives of democracy, that of providing equal opportunities to every member of society.

As its name indicates, the OU is an institution for open or nonresidential higher education. It offers courses and degree programs in seven subject areas, one of which is natural sciences. The only requirement to be met for admission in the OU is to be older than 18; certification of formal education is not a prerequisite. Neither does the OU require the students to reside or attend courses on a campus; freedom of location, time and pace of learning is an essential component of OU's philosophy (e.g., Crombag et al., 1979; Kirschner et al., 1993). To ensure that learning can take place wherever each individual student lives or works, instruction is carried out via printed materials designed didactically in such a way that they allow the learner to obtain continuous feedback about his or her progress. In some cases, written notes alone do not suffice and must therefore be complemented by other materials. This is particularly true in fields like the natural sciences, where exposure of the students to experimental work (laboratory or field experiments and measurements) is necessary. By using modern multimedia technologies (e.g., videodiscs), this need can be partially, and sometimes even largely, satisfied.

Like all Dutch universities, the OU is subsidized by the state. The cost associated with a full-year course load is similar to the tuition per academic year at regular universities. For Dutch nationals the fee for a 100-h course is DFl. 280 (1994, $\approx \$140$), including the registration fee, tuition, advising, and examination.

At the foundation of the OU, it was predicted that its enrollment would eventually reach between 20 000 and 30 000 students. In 1991 however, total enrollments at the OU had already climbed to >60 000 students. This number included 4162 students in the Department of Natural Sciences, 322 of which intended to complete a degree program. Approximately 40% of the OU students lack the formal certification (i.e., high school degree or equivalent) needed to get access to traditional institutions of higher education. This high percentage suggests that the OU successfully fulfills its role of providing a second chance to get access to higher education. The fact that the majority (60%) of the students are well-educated indicates that the OU also contributes to satisfying the educational needs of a rapidly changing society, in which time and geographical constraints may prevent people who wish to deepen their knowledge or expertise in a given field from enrolling at regular universities.

In the long run, it is clear that the answer to the question of whether or not the OU is successfully achieving its objectives will depend on how its alumni fare on the job market. The first group of OU students to graduate will do so in 1993. The fact that several of the natural sciences students have already received job offers even before they complete their degrees is an encouraging sign. It shows, indeed that the degree programs of the OU are perceived favorably by the general public.

The OU differs from traditional institutions of higher education not only because of the groups of citizens it targets and because of its underlying philosophy of open distance education. Multidisciplinarity also plays an important role in the conception of the educational programs of this young university; the OU has taken up the challenge of developing courses that are innovative in terms of both contents and format.

DISTANCE EDUCATION IN THE NATURAL SCIENCES

Philosophy

The natural sciences involve discovering, describing and understanding the living- and non-living components of nature. The methodology used by scientists in this field are based on a combination of observation, experimentation, and scientific reasoning, all of which are founded on a strong basis of physics, chemistry, biology, and geology. Rather than strictly disciplinar (e.g., physical) outlooks on nature, however, approaches that are at the frontier of various disciplines have generally become the rule. The study of soils is no exception in this respect.

The natural sciences curriculum at the OU attempts to integrate physics, chemistry, life sciences, and earth sciences. Designed with a problem-oriented perspective centered on the themes of the environment, on one hand, and of nutrition and toxicology, on the other, the degree programs offered by the OU involve new combinations of the traditional basic disciplines of the natural sciences; natural phenomena are described and analyzed both

Table 14-1. Professional specialties of the staff members in the Department of Natural Sciences.

Specializations	no.	Specializations	no.
Physics	2	Physical geography	1
Chemistry	2	Soil science	1
Biology	2	Nutrition	1
Geology	2	Health science	1
Climate physics	1	Environmental sciences	1
Biochemistry	1	Human geography	1
Pharmacology	1	Enviromental management	1
Toxicology	1	Political science	1
Microbiology	1		

at the molecular level and in relation to the system(s) of which they form a part.

The degree programs of the OU also try to integrate natural sciences with social sciences by linking the understanding of natural phenomena with an analysis of problems of policy and management. This deliberate bias toward multidisciplinarity is also reflected in the wide variety existing in the professional specialties of the staff members in OU's Department of Natural Sciences (Table 14-1).

Practical exercises and laboratory experimentation constitute an essential part of traditional science education. They have to be deemphasized, however, in the context of the OU to remain in keeping with its philosophy (based on freedom of location, time, and pace of study). They are nevertheless not eliminated altogether. The students have to carry out two face-to-face (student–teacher) laboratories to acquire the skills of observation and experimentation (Kirschner et al., 1993). These multidisciplinary, problem-oriented laboratories take place during holiday periods in laboratory facilities at regular universities, and include short guest lectures by professors from different disciplines and institutions. In addition to these (limited) laboratory sessions, extensive use is made of novel learning technologies (e.g., interactive videodiscs and compact discs), in combination with more traditional didactic techniques (e.g., simulations and case-studies). These various tools, which have been shown to assist the learning process efficiently (e.g., Kirschner, 1991), are used in conjunction with the laboratory sessions as didactic methods for learning and practicing the activities that constitute the profession of natural scientist.

In the second half of the study program, 800 h of internship in a research institution are compulsory. Supervision of this internship by a researcher affiliated with this institution, and by a staff member of the OU, guarantees the quality of the work. An internship offers the opportunity of a unique learning experience in the design, experimentation, and reporting stages of a research project. At the same time, it affords a unique opportunity for close interactions between students and potential future employers (Daniels et al., 1992).

Table 14-2. Curriculum of the Department of Natural Sciences.

Full academic (M.Sc.) degree programs (5400 h study load)	Short academic degree programs (1100-1500 h study load)	Courses (50-200 h study load)

Environmental Sciences
Environmental Policy
  & Management
Nutrition & Toxicology
	Applied Ecology
	Environment & Chemistry
	Geosystems
	Ecotoxicology
	Environmental Management
	Biotechnology
	International Issues in
	  Environmental Sciences
	  (under development)
		Self-contained units
		Three academic levels
		Multidisciplinary

The Curriculum

The Department of Natural Sciences at the OU offers degree programs in three different areas; (i) environmental sciences, (ii) environmental policy and management and (iii) nutrition and toxicology (Table 14-2). The final academic degree awarded by OU is the Dutch equivalent of the M.S. degree.

The courses that make up these programs are divided into three levels. First level (undergraduate) courses provide students with a body of general, basic knowledge in given areas. Second and third level courses allow the students to improve their theoretical and methodological knowledge and skills, and to integrate various disciplinary viewpoints. In third level courses, self-discovery learning and problem-solving are emphasized.

Each degree program comprises ≈ 40 courses and a final research internship. The courses are designed in such a way that they can be studied more or less independently of each other. Because of this modular structure, multiple combinations of courses, corresponding to individual needs or interests, can be elaborated. In some cases, to satisfy the learning needs and upgrade the expertise of already well-qualified people, short academic degree programs are developed, focusing on one particular aspect of the full academic degree program (see Table 14-2).

Stages in Course Development

The first stage in the development of a new course at the OU (Sloep et al., 1993) involves, collectively, all the staff members of the department that is planning to offer the course (Table 14-3). A number of ideas are proposed, sometimes pointing to very different directions. A staff member, designated as the future course team manager, is then responsible for drafting a preliminary outline of the course. Already during this initial stage, contacts are established with external professional specialists and with educational technologists of OU's Centre of Educational Production. These contacts are a crucial component of the course development procedure. Even though staff members sometimes end up writing significant portions of the course materials, their key task is to integrate the external professional

Table 14-3. The course development procedure at the Department of Natural Sciences.

Stage	Task	Responsible	Time period, mo
1	Drafting preliminary outline of the course contents	OU† staff members	3
2	Drafting of course plan	OU staff member(s) + et‡, with advise of external specialists	3
3	Instruction of external specialists	ctm† + et‡	1
	Writing of first version	External specialists (+ ctm)	5
	Evaluation	ctm + et + external referee	2
	Writing of revised versions	External specialists (+ ctm)	5
4	Field trial of course material	et + ctm	3
	Finalizing course	ctm	5
5	Technical production	OU publishing department	3

† ctm, Open University (OU) course team manager (scientific staff member).
‡ et, OU educational technologist.

knowledge with the didactic expertise available within the OU. This approach permits state-of-the-art scientific and technical information to be made readily accessible to the targeted audience.

Drafting of the detailed course plan (Stage 2) involves the close collaboration of all the members of the course team, which guarantees an optimum tuning of individual contributions. After the course plan is approved by the team, the actual writing (Stage 3) begins, closely associated with the development of nonwritten materials. Once the initial work on these written and nonwritten materials is completed, they are analyzed and evaluated by students belonging to the group targeted by the course. These students provide comments on the content and format of the course and indicate major bottlenecks. Based on these comments, a final version of the course materials is elaborated. Since knowledge, in many fields, is changing rapidly, each course has to be partially or, in some cases, entirely revised every 5 to 7 yr.

In addition to course-specific testing, the OU also evaluates the experiences of students with new media technologies in the natural sciences. Extensive research has been carried out, for example, on the anticipated and actual objectives of natural sciences didactic exercises (Kirschner, 1991). The results of this research are used in the design of new types of exercises relying on electronic media and of multidisciplinary face-to-face laboratories.

CHANGES IN SOIL SCIENCE EDUCATION IN THE NETHERLANDS

In the Netherlands, soil science education at the university level has changed during the last decade. From being mainly qualitative, oriented toward soil genesis (e.g., particularly at the University of Utrecht) and toward assessments of land suitability for agriculture (e.g., at the Wageningen Agricultural University), soil science itself has evolved during the years, in the direction of a more quantitative, problem-solving approach. This shift has been evinced by progressively increasing emphasis placed on quantitative land evaluation, quantification of the spatial variability of soil characteristics and process-oriented modeling of soil processes. In recent years, soil science graduates have also more and more frequently found employment as members of teams concerned with environmental decision making. As a result of these trends, soil science education programs have increasingly had to try to prepare students in such a way that they would be able to use their scientific knowledge to elaborate quantitatively sound policy and management decision, for example in probabilistic risk assessments (Bouma, 1989; Montagne, 1987).

This evolution has been paralleled by a profound change of the concept of *soil*. Traditionally, the latter was defined as the natural medium supporting the growth of plants, with a lower limit corresponding to the maximum depth of penetration of plant roots or biological activity (e.g., Bates & Jackson, 1980). In recent years, the general public has become acutely aware that soils are far more than just a resource or a support for agriculture; among many other roles, they also serve as sinks or even, in some cases, threaten to become chemical time bombs (Stigliani et al., 1991). As a result, the concept of *soil* has changed, becoming much broader, particularly in the environmental debate and in discussions about soil protection policy (e.g., Blum, 1993). From this enlarged viewpoint, *soil* is referred to as the upper part of the earth's surface, influenced by human and ecological activities. The depth of the soil can be quite variable and depends on the lower limits of the influences of the social and ecological functions of the soil.

The Dutch soil protection act of 1986 mentions a number of functions for soils, e.g., support for buildings and infrastructure, production of crops and food, role as resource, and the ecological and aesthetical role of soils. Besides these various functions, the Council of Europe (Blum, 1993) recognizes two additional ones: the function of the soil as cultural heritage and as a biological habitat and gene reserve. A clear example of the enlarged concept of soil is related to the function of the soil as drinking water reservoir. In the Netherlands, the present lower limits of the *soil* with respect to this particular function can be as deep as 200 m.

THE COURSE SOIL AND ENVIRONMENT

Target Groups and Objectives

The course entitled *Soil and Environment* is primarily aimed at people involved in soils-related policy decisions, either as members or as leaders of decision-making teams. The course is being developed to meet their apparent needs for knowledge on soils and soil functions, within the context of the environmental policy of the Netherlands.

After studying the course, the students are expected to be able to judge both the (soil) scientific background of soil quality criteria or legislations, and to understand the socioeconomic and political context in which soil quality standards are established. The main learning objectives therefore are to gain insight in the functioning of the soil system within the general framework of Dutch ecosystems, and to get a good grasp of the various considerations on which soil quality policy is based. The course has the objective to introduce the students to the problem-solving approach of soil environmental research.

The course *Soil and Environment* is, in the classification of the OU, a third or upper level course, with a study load of 150 h. Its prerequisites are a sound background in the natural sciences.

Course Contents

When designing soil science curricula in a university environment, one should as much as possible attempt to respond to the needs expressed outside academe (e.g., Montagne, 1987). In the case of the environmental courses offered by the OU, the targeted publics of *upgraders* and *updaters* are mainly working in nonacademic sectors, e.g., in private companies or in governmental agencies. They are as a rule more interested in the problem-solving approach of soil environmental research than in the traditional, descriptive approach. This determines the direction to take in establishing the contents of courses at the OU.

In the course *Soil and Environment*, the above-mentioned enlarged concept of soil is used throughout. The focus of the course is on soils, soil functions, and soil policy within the Netherlands. Because of the high population density in this country (442 inhabitants per square kilometer in 1990), the environmental debate at the local and regional scales tends to be dominated by issues related to soil quality and to spatially-oriented soil functions. In the Netherlands, there are hardly any areas without conflicts on land use. Via targeted policies, however, the (central) Dutch government tries to achieve the multifunctionality of all soils, in order to assure sustainable use. A soil is termed *multifunctional* if its quality is such that the soil is able to effectively fulfill a number of functions simultaneously. In practice, both local and national authorities have to decide in each individual case to give priority to one function at the cost of other soil functions. In the course *Soil and Environment*, the students are confronted with these kinds of decisions in policy making.

The first block of the course describes the major properties of the soil and the basic processes that take place within it. This block provides the background necessary for understanding soil environmental research. Particular emphasis is placed on the spatial interactions of soil processes and on the behavior of water, nutrients, and chemicals in the soil.

The central theme of the course, the variety of functions of the soil, is dealt within the second block (Table 14-4). Each function is discussed on the basis of examples from the Netherlands. A large number of different uses of soils are discussed, along with their consequences for the environment. At the end of this block, decision-making processes relative to soil usages are analyzed in detail. In particular, the normative, administrative

Table 14-4. Contents of the block on *Soil Functions* of the course *Soil and Environment*.

Study unit†	Chapters in textbook	Cases in interactive video program
1	Soils in natural ecosystem	
2	The soil archive	
3	Bearer of buildings and infrastructure	
4	Supplier of resources	
5	Waste depository	
6		Point source soil pollution: galvanic factory premises
7	Producer of agricultural products	
8	Water reservoir	
9		Nonpoint source soil pollution: nitrate leaching by slurry application
10	Soil protection	
11	Multifunctionality of soils	
12		Soil protection: regional planning of soil protection areas

† Study-unit in course book Part 2.

and legal aspects of regional planning and (local) environmental policy are dealt with.

Three cases of soil environmental problems, based on recently published research form an integral part of the soil functions block (Table 14-4). These cases are developed to teach to the students the skills of problem-solving in environmental soil issues. This is done practically with the help of an interactive video program (see section below). At the same time, these three cases enable the students to become acquainted with the decision-making process involved in resolving conflicts on soil functions.

DIDACTIC TECHNIQUES, NEW TECHNOLOGIES

In designing a course for the OU, five groups of factors interact: (i) the nature of the objectives the course is meant to address, (ii) the type of students the course is targetting, (ii) the subject matter, (iv) the didactic approach of the OU, and (v) constraining conditions like time and money. Careful consideration of these often conflicting factors is required before one can make a choice among a number of didactic alternatives available for the development of a given course. These alternatives generally differ in their didactic functionality (e.g., in the amount of guidance made available to students).

In the case of the course *Soil and Environment*, a combination of didactic alternatives were chosen, rather than a single one. Throughout the course the amount of guidance provided within the learning material (course book) decreases in parallel with a gradual increase of the freedom of study. A schematic diagram of the didactic design is given in Fig. 14-1.

The first block of the course has to introduce students in a rapid and effective mannerto the basic principles of soil science, a prerequisite to

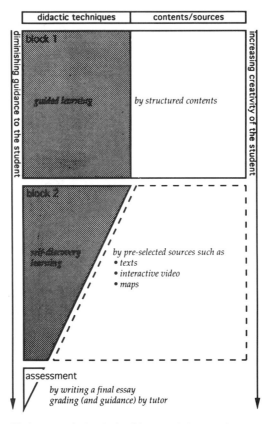

Fig. 14-1. Relationship between the level of guidance and the use of particular didactic techniques throughout the course *Soil and Environment*.

understand the material on soil functions in Block 2. The educational content of Block 1 is presented in study units (Table 14-5) and offered as course book Part 1.

The study-unit method is designed to enable the students to gain knowledge by self-study in an effective way. The contents are divided into units of an average length of 25 pages (4 h or a one-shift study load). A variety of techniques gives the student extensive didactic guidance. The didactic techniques, integrated in the educational content of the text, consist of carefully and explicitly stated learning objectives, key words, margin texts, in-text questions, tasks, and self-assessment questions (see Table 14-6).

The second block of the course, on soil functions, offers various sources of information by means of workbook-source material. This block is developed to teach the students the skills and attitudes for problem-solving in soil environmental research. The use of different types of source materials, along with the course book, facilitates self-discovery learning. The study-units in the course book Part 2 provide less didactic guidance, compared with the study-units of course book Part 1 (see Table 14-5). Emphasis is put in-

Table 14-5. Didactic structure of the course *Soil and Environment* for open distance education.

Component	Contents	Didactic technique	Study load
Course book Part 1	Properties of the soil; basic soil processes	Study-units with stated learning objectives, key words, margin texts, in-text questions and self-assessment questions.	65 h
Course book Part 2	Introduction to and explanation of the textbook chapters and the cases of the interactive video	Workbook with stated learning objectives, tasks and self-assessment questions	57 h
Textbook	Soil functions and the decision-making process	Textbook with key words	
Maps	Spatial information	Selected fragments	
Interactive video program	Information on three cases of soil environmental issues in the Netherlands	Computer-aided learning with preselected information and in-program guidance	
Assessment	Essays on cases of the interactive video program	Grading (and guidance) by tutor	28 h

creasingly on application of knowledge and on decision making. After an introduction and a list of learning objectives in course book Part 2, students are guided by means of tasks towards other sources of information. These additional information sources are a textbook with chapters on soil func-

Table 14-6. Explanation of the didactic techniques used in course book Part 1 (study unit variant).

Didactic technique	Explanation
Learning objectives	Each study-unit begins with the learning objectives. These objectives are statements of what the study should be capable of doing after studying the material in the unit.
Key words	New terms and concepts are known as key words. They are indicated in both the text and in the left hand margin. The key words structure the contents and give guidance in studying the text and retrieving information.
Margin texts	Margin texts function as informal comments and hints of the teacher.
In-text questions	In-text questions are followed immediately by an answer. These are meant to make the student pause and give the opportunity to recapitulate the above given information.
Tasks	Tasks are aimed to make the student use important concepts and put his or her newly acquired knowledge to use through a practical application.
Self assessment questions	Self-assessment questions appear at the end of each study unit. These questions enable the student to check whether the important concepts in the text are understood. Each question is related to the learning objectives of the study-unit.

tions and soil policy (Ivens and Lansu, 1993), (thematic) maps, and an interactive video program with various information materials describing real-life cases of environmental problems relative to soils.

Updaters and upgraders can choose the relevant study-units on the basis of their specific learning objectives. For professional updaters, the textbook provides a state-of-the-art coverage of environmental soil science in the context of Dutch soil policy.

The cases serve the dual role of problem-solving exercises and grading tools. The information on the case problems forms part of the block on soil functions and soil policy. The specific contents of the cases will be elaborated on. The students have to write an essay on each case. The first case, on source point pollution, is guided didactically by self-assessment of the written assignment. Students get acquainted with problem-solving strategies to disentangle the multifaceted information available on the item. The second case, on nonpoint source pollution, and the third case, on regional planning of soil protection areas, serve as grading assessments. A correspondence tutor is available for reviewing and grading the essays. One of the prerequisites to obtaining satisfactory grades in these essays is to have a good grasp of the basic concepts of soil science (at the level of Block 1).

PRACTICAL EXERCISES ON INTERACTIVE VIDEO DISC

In a professional context, soil environmental issues are analyzed and problems are solved by using a wide variety of information sources. The professional scientist extracts the data needed by reading reports and scientific papers, analyzing maps and remote sensing images, interviewing experts, using computer models and geographical information systems, and making field trips. In regular education programs, these information-gathering skills are taught in workshops or via research projects with direct student-tutor contacts. In the absence of face-to-face and time-constrained practical exercises, as is the case in distance education, the only way to simulate the same activities and situations as in the professional context seemed to be by using an interactive video program. An educational interactive video program (IV) consists of a video disc controlled by the student via a computer-assisted learning program. It is an effective didactic tool to offer variable sources of preselected information (Hannaway et al., 1988), to analyze spatial information (Maguire, 1989) and to enable students to be exposed to practical exercises such as laboratories and field trips.

As part of the course *Soil and Environment*, an interactive video program is designed after the typical work desk of a professional. The interactive video program offers to the students written information (report fragments) and audio-visual information (interviews, field trips, maps, figures, and pictures) concerning the three cases on environmental soil issues. Besides these information sources, interactive use is possible of preprocessed results of geographical information systems (GIS) and soil process models.

In each case study, the students are placed in the role of a professional are assigned specific tasks.

All three cases are situated in one physical geographical unit, the eastern part of the province of North-Brabant. In this region, agriculture, nature, and industry make conflicting claims on the use of soils. The first case deals with a point source pollution on the premises of a former galvanic factory (Staritsky et al., 1992). This is an example of deterioration of soil quality at a local level. The students have to evaluate a number of clean-up strategies. The second case, on nonpoint source pollution, deals with the threat to drinking water supplies created by NO_3 leaching from agricultural lands (Kragt & de Vries, 1990). The students are asked to consider the regional effects of several scenarios for restricting slurry application, with the help of GIS and a NO_3 leaching model (RENLEM). In the final case, the students are confronted with the regional planning of soil protection areas and have to evaluate the consequences, the effectiveness and the feasibility of such set-aside areas (Eweg, 1994; Bevers et al., 1991).

Each case is divided in several topics and each topic consists of selected, multifaceted, sources of information. The sources can be browsed directly, or with the help of an electronic tutor, with guidance on demand. The sessions involving the interactive video program are estimated to last ≈ 4 h for each case. They take place at one of the regional study centers of the OU where video disc players and multimedia computers are available. At the current speed of technological innovations, it is expected that distance students will be able to use these kinds of programs at home in the very near future.

A CHALLENGE FOR TRADITIONAL UNIVERSITIES?

At present, one is wittnessing an almost unbridled growth of educational programs dealing with environmental sciences, as well as the reorientation of existing programs in that same direction. Many of these new or reformed education offerings are limited by their reliance on traditional and disciplinary concepts and methodologies. By contrast, the multidisciplinary and problem-oriented approach to environmental issues, described above, opens new horizons. Although it seems particularly adapted to open distance education, the central theme of soil functions could also be adopted in regular soil science education at traditional universities.

The design and development of an interactive video program requires an extensive investment of time and money in professional, educational and technological expertise. Both expertise and sufficient financial resources may be lacking in traditional universities to carry out this type of endeavor. Nevertheless, it is likely that the interactive video program described above, or comparable ones, would also be useful in traditional education programs based on direct teacher-student contacts. An individual practical session with the interactive video program teaches to students the skills of soil environmental research. Such an individual session assures an effective preparation before starting a research project, without any time investment on the part of the tutor.

REFERENCES

Bates, R.L., and J.A. Jackson (ed.). 1960. Glossary of geology. Am. Geol. Inst., Falls Church, VA.

Bevers, A.M., A.B.M. Boezeman, and R. Siebinga. 1991. Soil protection areas, dead and buried? Milieu 4:109–114.

Blum, W.E.H. 1993. Soil protection concept of the Council of Europe and integrated soil research. Soil Environ. 1:37–47.

Bouma, J. 1989. Using soil survey data for quantitative land evaluation. Adv. Soil Sci. 9:177–213.

Crombag, H.F., T.M. Chang, K.D.J.M. van der Drift, and J.M. Moonen. 1979. Educational materials for the open university: Functions and costs. (In Dutch.) Dep. of Education and Sciences, The Hague, The Netherlands.

Daniels, W.L., J.R. McKenna, and J.C. Parker. 1992. Development of a B.S. degree program in environmental science. J. Nat. Resour. Life Sci. Educ. 21:70–74.

Eweg, H.P.A. 1994. Computer supported reconnaissance planning. Wageningse Ruimtelijke Studies 11. Agricultural Univ., Wageningen, the Netherlands.

Hannaway, D.B., P.J. Ballerstedt, P.E. Shuler, D.P. Cunnell, and D.N. Osterman. 1988. An interactive videodisc module for forage quality and testing instruction. J. Agron. Educ. 17:119–121.

Ivens, W.P.M.F., and A.L.E. Lansu (ed.). 1993. Soil and environment in the Netherlands. (In Dutch.) Wolters-Noordhoff. Open Univ., Groningen, Heerlen, the Netherlands.

Kirschner, P.A. 1991. Practicals in higher science education. H. Lemma, Ultrecht, the Netherlands.

Kirschner, P.A., M.A.M. Meester, E. Middelbeek, and H. Hermans. 1993. Agreement between student expectations, experiences and actual objectives of practicals in the natural sciences at the Open University of The Netherlands. Int. J. Sci. Educ. 15(2):175–197.

Kragt, J.F., and W. de Vries. 1990. Application of RENLEM on seven vulnerable groundwater protection areas. Staring Centre Rep. 38.3. Winand Staring Centre, Wageningen, the Netherlands.

Maguire, D.J. 1989. The Domesday interactive videodisc system in geography teaching. J. Geogr. Higher Educ. 13(1):55–68.

Montagne, C. 1987. A core curriculum for soil science majors. J. Agron. Educ. 16:14–16.

Sloep, P.B., M.C.E. van Dam-Mieras, and W.H. de Jau. 1993. Environmental issues at a distance. p. 35–41. *In* M.F. Ramalhoto (ed.) Proc. 2nd European forum for continuing engineering education, Lisbon, Portugal.

Staritsky, I.G., P.H.M. Sloot, and A. Stein. 1992. Spatial variability and sampling of cyanide polluted soil on former galvanic factory premises. Water Air Soil Pollut. 61:1–16.

Stigliani, W.M., P. Doelman, W. Salomons, K. Schulin, G.R.B. Smidt, and S.E.A.T.M. van der Zee. 1991. Chemical time bombs: Predicting the unpredictable. Environment 4:4–9, 26–30.

15 Fostering Learner Self-Direction in Soil Science Graduate Courses: A New Paradigm

Philippe Baveye

Cornell University
Ithaca, New York

ABSTRACT

The learning projects carried out by soil scientists after they leave graduate school are predominantly self-planned and autonomous. Yet colleges and universities are not preparing their students for this kind of learning, different in many ways from that occurring in formal settings in response to teacher-controlled instruction. The abundant literature that has recently been published on autodidactic learning suggests that teachers can play an active role in boosting their students' competence in this respect. After reviewing this literature, I describe a one-semester graduate soil physics course designed to make students familiar with the steps involved in typical self-directed learning events. This course consists of classroom lectures, where the instructor tries to make as apparent as possible his own learning process, and of weekly one-on-one tutorial sessions in which the students are placed in situations resembling as closely as possible those they will encounter in their future professional activities. Qualitative observations on my experience with this format since 1985 are analyzed and discussed in detail, particularly in terms of difficulties associated with the students' occasionally negative reactions. A number of perspectives for improving the course format are briefly outlined.

Every man who rises above the common level has received two educations: the first from his teachers; the second, more personal and important, from himself.

E. Gibbon (1796)

The truth is that even those who enjoy to the greatest extent the advantages of what is called a regular education must be their own instructors as to the greater portion of what they acquire, if they are ever to advance beyond the elements of learning. What they learn at schools and colleges is comparatively of small value, unless their own afterreading and study improve those advantages.

G. Craik (1830)

Copyright © 1994 Soil Science Society of America, 677 S. Segoe Rd., Madison, WI 53711, USA. *Soil Science Education: Philosophy and Perspectives.* SSSA Special Publication no. 37.

In 1895, at the age of 21, Winston Churchill came to realize that, in his words, "scarcely anything material or established which I was brought up to believe was permanent and vital, had lasted. Everything I was sure or taught to be sure was impossible, had happened" (Churchill, 1972). Today, almost 100 yr later, the pace of social, economical, and political changes had not abated, as the recent events in eastern Europe and in the former Soviet Union amply demonstrate. Advances in science and technology occur at an equally breathtaking rate; it has been recently estimated that the amount of information in the world doubles every 4 to 7 yr (Apps, 1988; Gayle, 1990) and that half of what most professionals know when they finish their formal training is outdated in < 5 yr. Not only is there considerably more information than ever before, but links with technology have made its storage, transmission, and access much easier (Merriam & Caffarella, 1991). In this respct as in many others, developments that yesterday were in the realm of science fiction are now taken for granted (Candy, 1991).

Widespread adoption of some of these new technologies (e.g., video equipment, satellite transmission, and multimedia personal computers) has profoundly expanded the limits of the traditional classroom. In contrast, the conception of education itself does not seem to have evolved noticeably. At the margins, in experimental programs and in some areas where change has led to profound problems such as mass unemployment or extreme racial tensions, educational providers have been willing in recent years to envisage drastic reforms (see e.g., Halterman, 1983; Bawden & Valentine, 1984; Bauer, 1985; Bawden, 1988; Rohfeld, 1991). However, in most mainstream institutions, in schools, colleges, universities, and even in adult education organizations, it is undeniable that it is "still very much a matter of business as usual, and [that] there is little more than a cursory nod in the direction of equipping people for a rapidly changing and uncertain future" (Candy, 1991, p. 51).

This slow evolution of the mainstream views on education may be due partially to the fact that having, in the vast majority of cases, not received any training in the philosophy of education or the theories of cognition, college and university professors tend to replicate the model they have experienced during their own studies (Reetz, 1972; Dunleavy, 1986; Eble, 1988). As a result of this *imitative teaching* pattern (Cross, 1991), encouraged in many ways by higher education's current (research-oriented) reward system, the views on education of most college professors are, by and large, identical to those advocated > 150 yr ago. Now, as it was then, learning is generally equated to *being taught* (Knowles, 1975, 1984). Kidd (1973) observes that it is as if "education must be carried out in a rectangular room, and that learning only happens where they are [...] students and one teacher." Furthermore, like 150 yr ago, teachers pay very little attention to the actual learning that takes place in and outside the classroom (e.g., Bok, 1986, p. 153). A recent survey (Cross, 1991) reveals that relatively few college teachers see students as growing, developing learners and that only 26% of the science teachers are willing to consider as an essential teaching goal the preparation of the students for the learning that they will have to carry out during the rest of their lives.

In the last few decades, numerous educators (Jackson, 1986; Resnick, 1987) have argued that, in order to adequately meet increasingly diverse educational demands and to prepare people effectively for a rapidly changing world, a radical rethinking is required of our views on education and of the organization of formal educational systems at all levels, from kindergarten to doctoral programs (e.g., Naisbitt & Aburdene, 1985; Jones & Cooper, 1980; Guglielmino et al., 1987; Martin, 1991). The purpose of schooling can no longer be to simply transmit fixed bodies of information (Bok, 1986). As Knowles (1975) puts it, "In the civilization of our forefathers it may have been possible for people to learn in their youthful years most of what they would need to know for the rest of their life, but this is no longer true. [..] When a person leaves schooling, he or she must not only have a foundation of knowledge acquired in the course of learning to inquire but, more importantly, also have the ability to go on acquiring new knowledge easily and skillfully the rest of his or her life. [..] Education—or, even better, learning—must now be defined as a lifelong process."

These views have been echoed occasionally in the writings of soil scientists. Low (1970), for example, emphasizes the fact that "a scientist's education should not stop with the Ph.D. degree. If he is successful, he will have to continue the learning process the rest of his life." In a similar vein, Nielsen (1970) argues that an integral part of the question of how to teach soil physics courses is "how a student can be taught to use his imagination to continue to learn and to progress." Unfortunately, this important question has seldom been raised again, let alone answered, in the last 20 yr.

Addressing this question properly in the context of soil science education requires detailed, quantitative information on the way successful soil scientists continue to learn and acquire new knowledge, once they have left graduate school. Unfortunately, this type of information is lacking, in soil science as in most other fields (Candy & Crebert, 1991). Informal surveys that I have conducted over the years suggest, nevertheless, that only a minute minority of active soil scientists elects to learn by *being taught*. Whether in academia, in state and federal agencies, or in the private sector, most soil scientists apparently attempt to satisfy their learning needs themselves, via discussions with colleagues or by reading appropriate references on their own. This seems particularly true of those who routinely use computers in their work; few of them would even envisage to enroll in courses to learn to use particular softwares when it is so much more convenient and practical to read self-study guides, available in most bookstores and updated regularly.

The fact that this *autodidactic* option is so widely adopted seems due to a number of reasons, several of which are common to many adult learners. Scheduling, location, and financial constraints are indeed frequent deterrents to participation in traditional continuing education programs (e.g., Penland, 1978, 1979; Cross, 1981). A more distinctive reason, however, is that formal (classroom) instruction often proves extremely unwieldy and frustrating to individuals with very specific and relatively high-level learning needs

(Hiemstra & Sisco, 1990; and the vivid example in Gross, 1991, p. 208). Whether or not it is the most expedient and efficient way to acquire new knowledge (Low, 1970), audodidactic learning often is for active soil scientists the only viable option to do so.

The key question presently confronting graduate soil science education, therefore, is whether there is anything at all that can be done to prepare and equip students for their future autodidactic learning. As recently as 20 yr ago, educational research findings provided little help to address this issue. In 1973, R. Gross (1973) came to the grim conclusion that "[..] the literature of education is virtually devoid of studies of individual learning in its real-life context." Since then, fortunately, the research on self-directed learning and autodidaxy has experienced a "meteoric rise to prominence" (Candy, 1991). As a result, the body of literature in this field has greatly expanded. Admittedly, many aspects of self-direction, autonomy in learning and autodidaxy are still poorly understood. Nevertheless, there seems now to be enough of a conceptual foundation to begin to answer the question raised above. My intent in the following few pages is to describe what appears to be a first step made in this direction.

The present chapter is organized as follows. In the second section, immediately following this introduction, the results of studies on self-direction and autodidactic learning are reviewed critically. A number of excellent and thorough reviews of this field have been published recently (e.g., Hiemstra & Sisco, 1990; Smith, 1990; Candy, 1991; Merriam & Caffarella, 1991). Therefore, the coverage here focuses on the published results that bear direct relevance to fostering self-directed learning in graduate courses. Even in this restricted context, many of the very interesting philosophical developments of the last few years, like Candy's (1989) constructivist model or Tremblay and Theil's (1991) model of autodidaxy, are beyond the scope of the present chapter. This review serves as the conceptual background for the third section, in which I describe in detail the one-semester graduate soil physics course that has been designed specifically to enhance students' self-directed learning skills. In particular, the quintessential component of this graduate course, the individual weekly tutorial, is analyzed in detail. Some of the problems and difficulties encountered with the course format since 1985 are discussed in the fourth section. Finally, in the last part of this chapter, prospects for future developments of the tutorial format, as well as some of the aspects of the approach that are in need of further confirmatory research, are briefly outlined. It is hoped that the few thoughts presented in the following pages will be of some use not only to those involved in the teaching of soil science graduate courses but also to all scientists interested in reflecting about their own learning processes.

EDUCATIONAL SCIENCES BACKGROUND

Learner-Control Versus Autodidactic Learning

In recent years, self-direction in learning has progressively emerged as one of the most attractive and vital concepts in educational research and practice. This movement has resulted in many hundreds of masters' theses, doctoral dissertations, journal articles, research reports, conference papers, and books. The first contact with this huge and rapidly expanding amount of literature is intimidating and often confusing. One reason for this is the profusion of labels under which the field is referred to: autodidaxy, autonomous learning, independent learning, individual self-planned learning, learner-controlled directed instruction, learner-managed learning, nontraditional learning, open learning, participatory learning, peak learning, self-directed learning, self education, self-organized learning, self-planned learning, self-reliant learning, self-responsible learning, self-study, and self-teaching. As Candy (1987) notes, "this proliferation of terms would be difficult enough if they were all exact synonyms, but the problem is made worse by the fact that different authors use the same term to mean different things, and sometimes they use different terms to mean the same thing." As an example of the first situation, Moore (1973) identified at least four different meanings commonly encountered for the term *independent study*: correspondence courses; individualized, programmed instruction in a school setting; supervised reading programs in school; and out-of-school, part-time degree programs for adults.

Fortunately, significant efforts have been made in recent years to dissipate some of the confusion associated with this multiplicity of labels. Candy (1989, 1990, 1991), in particular, has proposed an appealing four-way classification, which will be adopted in the following. He suggested that the term *self-direction in learning*, as used in the literature, embraces dimensions of process (i.e., means) and product (i.e., goal), and that it refers to four distinct concepts: *self-direction* as a mode of organizing instruction in formal settings (learner-control); *self-direction* as the individual, noninstitutional pursuit of learning opportunities in the *natural societal setting* (autodidaxy); *self-direction* as the willingness and capacity to conduct one's own education (self-management); and *self-direction* as a personal attribute (personal autonomy).

To understand the operational limits of attempts to foster the capability for self-direction in learning, it is important to grasp the differences existing between the two *process* components of self-direction: learner-control and autodidaxy. To this end, it is useful to review briefly why and how these two components became the object of so much attention by educational researchers.

The last few decades have witnessed a growing realization that teacher-controlled education often nurtures the dependency of students upon teachers (e.g., Kahnweiler, 1991; Grow, 1991a). The learning orientation that is associated with this dependency has been occasionally termed *achievement-oriented* learning (e.g., Candy, 1991). It has to do with using whatever learning strategy is appropriate to attain high grades or to please the teacher. A genuine interest in the subject matter itself is not a prerequisite for this type of learning, nor is it always one of its outcomes. Competencies needed to succeed at it involve patient note taking, the ability to spot examination questions, and a good memory (Jackson, 1986; Cornwall, 1988). As Boud (1988, p. 35-36) points out, "there have been a number of notable studies over the years which have demonstrated that assessment methods and requirements probably have a greater influence on how and what students learn than any other single factor. This influence may well be of greater importance than the impact of teachers or teaching materials."

Over the years, various alternatives have been proposed in order to discourage the adoption by the students of a strictly achievement-oriented approach to learning. Several of these alternatives attempted to maintain complete control by the teachers during the instructional process. Elton and Laurillard (1979), for example, suggested the adoption of assessment tasks that require deep approaches to learning and thereby discourage students from using reproducing strategies. Most often, however, the approaches that were advocated amounted to surrendering to learners the control of certain aspects of the instructional situation (e.g., Hiemstra & Sisco, 1990). These approaches have raised considerable controversy (e.g., Knowles, 1975; Brookfield, 1986; Candy, 1987, 1990; Pratt, 1988; Merriam & Caffarella, 1991, p. 53). Nevertheless, the motivation for this partial or total surrender has come from research findings that strongly suggest that, except perhaps when learning is defined in conventional terms as the acquisition of a certain amount of factual information, learner-control leads to enhanced learning outcomes. Indeed, as noted by Candy (1991, p. 242), learner-control seems to entail collateral gains in curiosity, critical thinking, and information seeking behavior. Furthermore, the quality and retention of understandings appear to be enhanced when learners have the responsibility to *sort out* essential from inessential information.

A parallel development in the last 25 yr has been the progressive recognition of the crucial importance of autodidaxy in adult education. For all practical purposes, the scholarly study of autodidaxy can be traced back to the appearance in 1961 of a book by C.O. Houle (1961). In this essay, the author sketched the learning motives and activities of 22 continuing learners, who chose to pursue their learning at length and in depth, without institutional support or affiliation. Basic surveys have shown that there is a great deal of consistency among adult populations in terms of the learning projects that people conduct. For example, in a survey of adults of varying educational levels and socioeconomic status, Tough (1978) found that 90% of all adults conduct at least one learning project per year and that the source of the planning of these learning projects is as illustrated in Fig. 15-1. Operationally, Tough (1978) defined a learning project as an effort of at least 7-h duration; in fact the research found that the average project took 100 h. Learners were found to conduct five such projects a year, on average, spending almost 10 h per week on learning projects. For much of the learning, >80% of it, an *amateur* does the day-to-day planning (Fig. 15-1). Even more striking is the fact that in 73% of the cases, the *amateur* who does the planning is the learner himself.

Each of the four concepts of self-direction mentioned above (i.e., learner-control, autodidaxy, self-management, and personal autonomy) involves a great number of components (e.g., Candy, 1990, 1991, p. 242). Nevertheless, it is useful to think of each of these concepts as a lumped, one-

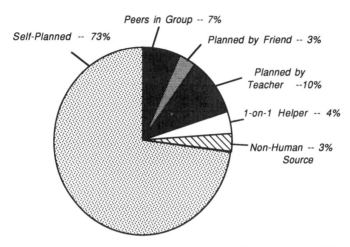

Fig. 15-1. Sources of planning for adults' learning projects (After Bonham, 1992).

dimensional continuum, where all the various components interact and mutually modify each other. In the case of learner-control (i.e., self-direction in formal settings), various instructional strategies could be placed at intervals along this continuum, to imply the differing balance of teacher-control and learner-control (Fig. 15-2). According to Candy (1991, p. 10), "at the far left of the continuum might come indoctrination (a), with almost total teacher-control and little room for learner-control at all. Then might come, in sequence, lectures (b), lessons (c), programmed instruction (d), individualized instruction (e), personalized instruction (f), interactive computer-managed learning (g), discovery learning (h), and so on, until finally the point is reached where learners have accepted almost all control over valued instructional functions." Candy (1991) refers to this point (i), at the far right-hand edge of the continuum, as *independent study*.

Autodidaxy also may be portrayed diagrammatically as falling along a continuum, where the self-instructional situations are distinguished not by the level of control of the learner, which in this case is total, but by the level of assistance sought. Indeed, even though initiative for a learning project rests firmly and indisputably with the autodidact, it is possible that he or

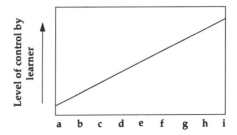

Fig. 15-2. A hypothetical learner-control continuum showing instructional strategies ranging from indoctrination (a) to independent study (i) (adapted from Candy, 1991).

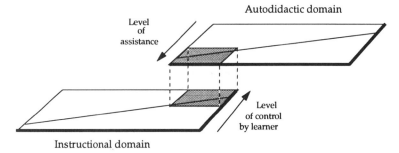

Fig. 15-3. Learner-control and autodidaxy as *laminated* domains (adapted from Candy, 1991).

she might make extensive use of a guide or a helper, or perhaps even more than one, to assist with a range of factors going from emotional encouragement, to the location and utilization of specific resources, to management of the learning process itself (Danis & Tremblay, 1985a,b).

At first sight, it seems that little distinguishes a situation of assisted autodidaxy from one of independent study. This viewpoint, if it were correct, would lead to the notion that there exists a single continuum extending all the way from a high degree of teacher-control to *pure* autonomous learnig or autodidaxy, where the learner receives no assistance of any kind. Candy (1991) argues forcefully that this is not the case, because the issue of *ownership* clearly differentiates independent study from assisted autodidaxy. At the point i in Fig. 15-2, there is still a residue, albeit small, of teacher control. Even though the instructor might have all but vanished, "the image of hierarchical power does not automatically disappear from the learner's mind" (Chené, 1983, p. 44). "Whether symbolically or otherwise, the instructor maintains some degree of control (and hence ownership) over the instructional transaction and, in the final analysis, independent study is still a technique of instruction" (Candy, 1991, p. 18; see also Bauer, 1985; Brocket & Hiemstra, 1985; Brookfield, 1985a, b). In the autodidactic domain, on the other hand, the learner is frequently not conscious of being a learner, much less a student, and hence the image of the instructor is not present to begin with (Thomas, 1967).

If, following Candy (1990, 1991), the *learner-control* and *autodidaxy* continua do not overlap, then how are they related? Candy (1990, 1991) suggests that they should be viewed as *laminated* or layered, as in Fig. 15-3. In the region where the autodidaxy domain is directly above the instructional domain, it may be difficult for a researcher or for some other outside observer to distinguish one situation from the other. Candy (1991) argues that only the participants can be certain about whether *ownership*, particularly ownership of the original questions that guide a learning endeavor, has been transferred to the learner, and even they may be unclear at times.

In Fig. 15-3, the autodidactic domain is pictured above the instructional domain, a configuration that may suggest that the autodidactic learning is viewed as somehow superior to the learning occurring in formal settings. Actually, opposite views are frequently expressed in the literature. Dickinson (1979, p. 4), for example, believes that learning in natural societal settings is "an inefficient way of learning which may even be harmful to the learner since no one is guiding the activity." Similarly, Candy (1991, p. 288) comments that, concerning any given phenomenon, there is usually a *preferred* or *correct* way of thinking, sanctioned by the canons of formal science, and that "in the case of self-directed learners, there may be no built-in or inherent mechanism to ensure that a learner does confront discrepancies between his or her present way of thinking and that which is sanctioned by the discipline or body of knowledge involved." Candy (1991) seems to consider this a serious disadvantage of autodidactic learning, compared with teacher-controlled instruction where it is far easier for the learners to absorb the *canon of formal science*. I believe, on the contrary, that this difficulty of separating autonomy in learning from autonomy in thinking probably results in one of the key advantages of autodidaxy. From this viewpoint, it is not a mere coincidence that self-education has been an important tool in the lives of prominent thinkers throughout the history of Western Civilization, like Socrates, Aristotle, Benjamin Franklin, Jean-Jacques Rousseau, George Green, and Albert Einstein, to name only a few.

In the conceptual framework illustrated schematically in Fig. 15-3, a *castastrophic* transition, a little like a phase transition in physical chemistry, is needed before the students can operate on their own in the autodidactic domain. In a later subsection, I shall be interested in the kinds of activities, within the instructional domain, that are thought to facilitate this transaction. Before I do so, however, it is useful to review briefly what is known about the learning patterns and the personality traits of autodidacts, as well as the various tools available to assess autodidactic learning readiness.

Learning Patterns of Autodidacts: Linear or Serendipitous?

Any person who more or less deliberately sets out to learn something on his or her own has to answer a series of questions that Biggs (1986, p. 143) summarizes as follows:

Motives	"What do I want?"
Goals	"What will it look like when I've got there?"
Task demands	"What do I need to get there?"
Context	"What resources have I got to use?"
	"What constraints must I contend with?"
Abilities	"What am I capable of doing"
Strategies	"Well, then. How do I go about it?"

How autodidacts answer these questions in practice has been a major area of inquiry in the last two decades. Until quite recently, it was assumed that the process of autodidactic learning was similar in nature to the *formal* learning process in which learners are taught by a teacher. Autodidacts were seen as planning and carrying out their learning activities in a linear, sequential pattern: establishing goals and objectives, locating resources, choosing learning strategies (Merriam & Caffarella, 1991). Within the last decade, however,

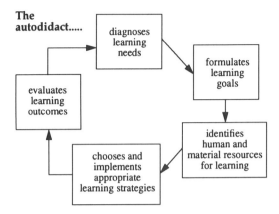

Fig. 15-4. Schematic diagram of the autodidactic process according to Knowles (1975).

alternative descriptions of autodidactic learning events have begun to emerge, which paint a very different picture.

Tough (1967, 1978, 1979) is apparently the first to have proposed a conceptualization of the ways people learn on their own. He did so on the basis of his background research and experience, interviews with learners and what he has termed logical analysis. His research resulted in a list of 13 steps that represent key decision-making points about choosing what, where and how to learn:

1. Deciding what detailed knowledge and skills to learn
2. Deciding the specific activities, methods, resources, or equipment for learning
3. Deciding where to learn
4. Setting specific deadlines or intermediate targets
5. Deciding when to begin a learning episode
6. Deciding the pace at which to proceed during a learning episode
7. Estimating the current level of one's knowledge and skill or one's progress in gaining the desired knowledge and skill
8. Detecting any factor that has been hindering learning or discovering inefficient aspects of the current procedures
9. Obtaining the desired resources or equipment or reaching the desired place or resource
10. Preparing or adapting a room (or certain furniture or equipment) for learning or arranging certain other physical conditions in preparation for learning
11. Saving or obtaining the money necessary for the use of certain human or nonhuman resources
12. Finding time for the learning
13. Taking steps to increase the motivation for certain learning episodes

Knowles (1975) has outlined a somewhat similar, but less detailed, sequence of steps. It is represented as a cyclical diagram in Fig. 15-4.

In essence, the process of learning on one's own, as conceptualized by Tough (1967, 1979) and Knowles (1975), is similar to the way in which teachers plan and carry out instruction in formal institutional settings. Whether this similitude is deep-rooted or whether it stems merely from the use of particular research methodologies is still open to debate, however. Until roughly the early eighties, the technique most commonly used to *capture* information about the autodidactic process was to present respondents with prepared lists of the function they may be expected to perform, such as the list proposed by Knowles (1975). Not unexpectedly, the respondents under these conditions tended to indicate that they did, indeed, undertake the sequence of tasks suggested to them. Candy (1991, p. 168) proposes three reasons for this; "First, research subjects are legendary for their desire and willingness to please the researcher, and will therefore commonly agree to propositions that they think the interviewer wants to hear. Second, there is often a difference between the actual way in which they should be accomplished and the *approved* or socially sanctioned way in which they should be accomplished, and respondents prefer to side with the *right* answer. Third, it is often the case that people have given little conscious thought to the steps involved in certain activities, and they simply respond to what appears to be a plausible sequence, since they are in no position to come up with anything better on the spot." With respect to the second of these reasons, one may add that it is likely that individuals who have been subjected to years of schooling would tend to recognize the linear pattern, typical of formal instruction, as the *correct* one.

In recent years, researchers have developed and used a variety of alternative methodologies to deal more adequately with the complex nature of self-directed learning outside institutional settings. Examples include reconstruction through interview, biographies and reflective essays, learning, journals and diaries, and recurrent interviews throughout the duration of a learning endeavor (e.g., Candy, 1991, p. 169-176). Using some of these techniques, several researchers have obtained observation data strongly suggesting that autodidacts do not necessarily follow a definite sequence of steps. Drawing on a study of 78 adults with less than a high school education, Spear and Mocker (1984) concluded that "self-directed learners, rather than preplanning their learning projects, tend to select a course from limited alternatives which happen to occur in their environment and which tend to structure their learning projects." They labeled this phenomenon the *organizing circumstance* and explained the process as follows (Merrian & Caffarella, 1991): (i) the triggering event for a learning project stems from a change in life circumstances; (ii) the changed circumstance provides an opportunity for learning; (iii) the structure, method, resources, and conditions for learning are directed by the circumstances; and (iv) learning sequences progress as the circumstances created in one episode become the circumstances for the next logical step.

In an exploratory study of 10 training and development personnel, Spear (1988, p. 212-213) found that the process of autodidactic learning could be reduced to seven principal components:

Knowledge
- $K_{(r)}$, Residual knowledge: knowledge the learner brings to the project as a residue from prior knowledge
- $K_{(a)}$, Acquired knowledge: knowledge acquired as part of the learning project

Action
- $A_{(d)}$, Directed action: action directed toward a known or specific end
- $A_{(e)}$, Exploratory: action that the learner chooses without knowing what the outcomes may be or with certainty that any useful outcome will ensue
- $A_{(f)}$, Fortuitous action: action that the learner takes for reasons not related to the learning project

Environment
- $E_{(c)}$, Consistent environment: includes both human and material elements that are regularly in place and generally accessible
- $E_{(f)}$, Fortuitous environment: provides for chance encounters that could not be expected or foreseen and yet affect the learner and the project

Spear (1988) proposed that each autodidactic learning event is composed of sets or clusters of these seven principal components. To illustrate this concept, following Merriam and Caffarella (1991), a learning cluster could be like the process described in the following (hypothetical) example. Elizabeth D., in the research center where she works, is part of an informal reading group ($E_{(c)}$) in soil physics. She decides that she would like to learn more about the theory of fractals, a subject she has already become somewhat familiar with during her graduate studies and through some independent reading ($K_{(r)}$). During a soil science symposium, she decides to attend ($A_{(d)}$) a session dealing with fractals, and finds most of the lectures to be interesting ($K_{(a)}$). At the end of one of these lectures, she decides some refreshments might be appropriate and she stops at a nearby coffee shop with a friend ($A_{(f)}$). By chance they encounter a group of conference participants discussing the talks on fractal theory that they have attended, and are invited to join their lively exchange ($A_{(e)}$).

The learning process described in this example may be looked at as a single cluster involving actions planned at the onset and some that are entirely the result of serendipity. In general, according to Spear (1988), autodidactic learning projects are composed of several such clusters and do not occur in linear fashion; one cluster of activities does not necessarily bear any relation to the next cluster. Rather, "information gathered through one set of activities is stored until it fits in with other ideas and resources on the same topic gleaned from one or more additional clusters of activities. Therefore, a successful self-directed learning project is one in which a person can engage in a sufficient number of relevant clusters of learning activities and then assemble these clusters into a coherent whole" (Merriam & Caffarella, 1991). Spear (1988, p. 217) concludes: "The learner is perhaps in greatest control when the assembling of the clusters begins and decisions are made regarding what knowledge is of most and least importance." As this process unfolds and develops, there may be concomitantly "a progressive refinement, or even wholesale reformulation of learning goals" (Spear, 1988).

Danis and Tremblay (1987) and Berger (1990) reach similar conclusions. The former authors made a detailed content analysis of the learning experiences of 10 autodidacts and conclude that these experiences do not conform to either a linear or a cyclical sequence, but rather that "the self-taught adults proceed in a heuristic manner within a learning approach which they organize around intentions, redefine and specify without following any predetermined patterns" (Danis & Tremblay, 1985a, p. 139). There is little evidence that the subjects of Berger's (1990) study (20 Caucasian males with no formal degrees beyond high school) had preplanned their self-directed learning activities. Most of them did not even make a conscious decision to start a learning project, but rather gradually became involved as a result of a particular triggering circumstance. Furthermore, they "constantly redefined their projects, changed course, and followed new paths of interest as they proceeded" (Berger, 1990, p. 176). These observations have permitted to shed new light on a group of phenomena examined earlier by Tremblay (1981) concerning the criteria that guide autodidacts in the choice of a resource. She observed that in 75% of the cases, the autodidacts questioned had not planned to use the resource consulted and that this had been chosen by chance or quite simply because it had come to hand.

The linear pattern of Tough (1967) and Knowles (1975), and the *nonalgorithmic syntax* (Tremblay & Theil, 1991) advocated by Spear (1988), Danis and Tremblay (1987), and Berger (1990) both have recently become viewed as extremes by a number of researchers. Long (1991a) argues that both patterns "may be true when applied to different people" but that, for most individuals, the truth is likely to reside somewhere between those two extremes. Merriam and Caffarella (1991, p. 50) concur with this viewpoint and suggest that a number of variables (e.g., individual motivation, knowledge of the content to be learned, and simple happenstance) determine the degree of linearity of autodidactic learning processes; "How the process of learning on one's own continues to evolve depends on the continuing interaction of these variables." The question, so far entirely neglected in the literature, of whether the autodidact has or should strive to gain control over this evolution, is important in practice and will be addressed in the following section.

Skills and Competencies of Autodidacts

Starting with Chickering (1964), various studies have tried to determine which attributes and competencies are either possessed by or desirable in autodidacts. The composite profile that can be assembled on the basis of the results of these studies indicates that the person capable of exercising control over the tasks to be mastered, and of working independently, would ideally (Candy, 1991, p. 130):

Be methodical and disciplined
Be logical and analytical
Be reflective and self-aware
Demonstrate curiosity, openness, and motivation
Be flexible
Be interdependent and interpersonally competent
Be persistent and responsible
Be venturesome and creative
Show confidence and have a positive self-concept
Have developed information-seeking and retrieval skills
Have knowledge about, and skill at, learning generally
Develop and use defensible criteria for evaluating learning

According to Candy (1991), the fact that similar composite lists abound in the literature is evidence of a definite cluster of competencies by which autodidacts might be recognized. One has to admit, however, that this list is far too general and idealistic to be of much guidance to autodidacts or to self-directed-learning facilitators. In this respect, the comprehensive yet concise list compiled by Skager (1979) appears much more useful. After an extensive review of the relevant literature, this author proposed seven types of attributes possessed by the *self-directed learner*. These are:

1. self acceptance or positive views about the self as learner, based on prior experience;
2. planfulness, which comprises the capacity to
   a. diagnose one's own learning needs,
   b. set appropriate goals, and
   c. select or devise effective learning strategies;
3. intrinsic motivation or willingness to persist in learning in the absence of immediate external rewards or punishments;
4. internalized evaluation or the ability to apply evidence to the qualitative regulation of one's own learning activity;
5. openness to experience and a willingness to engage in new activities because of curiosity or similar motives;
6. flexibility or willingness to explore new avenues of learning;
7. autonomy, or the ability to choose learning goals and means that may seem unimportant or even undesirable in the immediate social context.

Implicit in this portrait of the self-directed learner is the view that self-directedness is an intrinsic quality of the person rather than a characteristic or property of a learner in a given situation. In other words, it assumes that once a person attains autonomy as a learner in one domain, such as English or ornamental horticulture, he or she will automatically be able to learn autonomously in other unrelated fields, such as mechanical engineering or child psychology. Candy (1991), among others, vehemently opposes this view. According to him, "any person could vary in the degree of autonomy he or she exhibits from situation to situation." In other words, "this means that no self-directed learner can be equally competent across the range of all potential learning situations. While he or she may possess an extensive répertoire of strategic or general learning skills, each new domain will have its own domain-specific vocabulary of concepts that must be mastered before more advanced ideas can be tackled" (Candy, 1991, p. 304). Even if one considers plausible in general this concept of a limited transferability of competence from one learning situation to another, the question remains of the *threshold distance* between learning situations, beyond which this transferability decreases significantly. In the example above, the jump from ornamental horticulture to child psychology may indeed be a formidable one for many autodidactic learners. A person, however, having attained autonomy as a learner in, e.g., theoretical soil physics, may very well find it easy to learn autonomously in, e.g., the neighboring field of soil physical chemistry. Further research is needed on this issue before one can assess the range of validity of Candy's (1991) views on the *situationally variable* nature of autonomy in learning.

Lists of attributes, like the two presented above, are often used in another context; to characterize the profile of individuals with given learning styles (see, e.g., Witkin et al., 1977; Raven, 1992). A number of researchers have tried to find out if there is any correlation between learning styles and competencies of autodidacts. Their findings, however, are confusing and contradictory at best (Merriam & Caffarella, 1991), and provide little guidance at this point to those who want to foster self-direction in learning. As an example, Pratt (1984) argues, from a conceptual perspective, that people with tendencies toward field independence are more capable of self-directed learning. Similar views have been advocated by Even (1982). Brookfield (1986), in contrast, states that field dependence is more characteristic of self-directed learners. Brookfield's view is grounded in the idea that "self-directed learning is equated with the exhibition of critical reflection on the part of adults" and that the kinds of beliefs (such as the contextuality and relativity of knowledge) needed for this kind of thinking are most often seen in field-dependent vs. field-independent learners.

Assessing the Readiness for Audodidactic Learning

In an attempt to make the research on self-direction more quantitative, various theorists have sought to develop tests, instruments, and questionnaires that purport to measure aspects of learner autonomy. Before I analyze the various ways that have been proposed to foster self-direction in learning, it is interesting to briefly discuss if, and to what extent, these quantitative tests are useful.

Two of the best known instruments are Guglielmino's Self-Directed Learning Readiness Scale (SDLRS) (1977), and Oddi's Continuing Learning Inventory (OCLI) (Oddi, 1985, 1986). Both of them assume generalizability of competence of autonomous learners. Neither differentiates between *independent* continuing professional education within institutional settings (such as graduate schools) and autodidactic continuing education outside formal settings.

The SLDRS consists of a series of 58 statements like *I love to learn* or *No one but me is really responsible for what I learn.* In each case, the individual tested is asked to indicate on a five-point Likert scale the degree to which she or he feels that the statement is true for her or him. After scoring of the 58 answers, the SDLRS provides an assessment of *readiness* ranging from *high* to *low.*

Numerous studies have been completed using the SDLRS during the past decade. Guglielmino et al. (1987), for example, have shown that outstanding performers in jobs requiring a very high level of creativity or a very high degree of problem-solving skill had significantly higher SDLRS scores than others. In some cases, the use of the SDLRS has resulted in puzzling contradictions. For example, Guglielmino et al. (1987) and Adenuga (1991) observed that individuals who have completed higher levels of education tend to have higher SDLRS scores, while Long (1991b) concluded from a study of full-time and part-time college students that "there is no association between SDLRS and education achievement level, defined in terms either of years of schooling or quarter hours of college work completed."

Use of the SDLRS in research on self-direction in learning has not been without controversy. Major questions have been raised by Field (1989), among others, as to its basic validity and reliability. In particular, Field (1989) argued that what the SDLRS actually measures "is not readiness for self-directed learning, but does appear to be related to love and enthusiasm for learning." In the lively debate sparked by Field's (1989) criticisms, a general consensus seems to have formed that what the SDLRS actually measures is some form of *perceived readiness*. Whether there is "congruent or disjunction between adults' own judgments regarding the quality of their learning and that quality as measured by some external, objective standard" (Brookfield, 1985a) is still an open question.

The second instrument, the OCLI, was developed, in part, as a reaction to the SDLRS (West & Bentley, 1991). Instead of focusing directly on learning preferences, like the SDLRS, it attempts to identify clusters of personality characteristics found to relate to "initiative and persistence in learning over time through a variety of learning modes" (Oddi, 1985, p. 230). These clusters of personality dimensions were developed after carefully reviewing lists of attributes like the ones presented in the section above. As a result, the assessment provided by the OCLI may be less related to the perception of the individuals tested than is the case with the SDLRS. Practically, the OCLI consists of a self-reported scale with 24 items like *After I read a book or see a play or film, I talk to others to see what they think about it* or *I regularly read professional journals.* In each case, the individual tested is asked to record along a seven-point Likert scale the extent of his or her agreement. Like the SDLRS, the OCLI fails to distinguish between the various concepts of self-direction. In this respect, Candy (1991) argues that the OCLI may be more appropriate to the domain of learner-control than to that of audodidaxy while Six's (1987) results, on the contrary, suggest that "the scores on the OCLI have nothing to do with predicting student behavior in an instructional setting." Only very limited use has been made of the OCLI so far. It too, like the SDLRS, has received some criticism (e.g., Landers, 1989) even though the study on which it is based is generally considered to have been well controlled (Candy, 1991).

In a recent study (West & Bentley, 1991), both the SDLRS and the OCLI were administered to 810 teachers and administrators in 30 public schools (primary, middle, and high schools). One of the conclusions of this survey is that neither instrument appears very helpful in predicting those who would participate in a greater number of self-directed learning activities. This conclusion should not necessarily discourage the use of either the SDLRS or the OCLI, however. Neither one may be particularly effective as a screening tool or to monitor increases in self-direction over time. Nevertheless, to those who attempt to foster learners' self-direction, the SDLRS and the OCLI may still provide useful information on how the learners' perception evolves over time. Reference to the use of the SDLRS and the OCLI in this context will be made in a later section of the present chapter.

Fostering Autodidactic Learning Competence

Is autodidactic competence susceptible to educational interventions in formal settings? In the affirmative, what sort of educational interventions enhance the capability for autodidaxy? Answers to these critical questions have been quite varied in the last few decades. Collins (1988), for example, is emphatically opposed to educator intervention in self-directed learning, arguing that educators only intrude and actually erode further prospects for genuinely autonomous learning. Most authors, however, adopt a different viewpoint and generally seem to agree that, in the instructional domain (Fig. 15-3) at least, educators can definitely make a concrete contribution to the development of self-directed learning competence. In so far as the autodidactic domain is concerned, Resnick (1987) remarks that "the evidence developed [...] on the discontinuity between school and work [as learning environments] should make us suspicious of attempts to apply directly what we know about skills for learning in schools to the problems of fostering capabilities for learning outside school." Candy (1991) similarly argues that "there is [...] something incongruous about attempts to enhance the ability of learners to function outside the structures of formal institutions from within the institutions themselves." Even though the link between learner-control and autodidaxy is conceptually very tenuous and remains largely to be verified through detailed research, increased self-directed learning competence in the instructional domain is implicitly assumed by many educational theorists to lead naturally to autonomous behavior in the autodidactic domain. It is as if, by encouraging students to move along the learner-control continuum in the instructional domain (Fig. 15-3), educators were providing them enough momentum to make on their own the transition to the autodidactic domain, somewhat like fighter jets being propelled off the deck of an aircraft carrier and flying by themselves thereafter.

The literature suggests that there are two broad approaches to develop in people both the ability and the willingness to take charge of their own learning processes.

The first might be referred to as *direct instructional intervention* (Wang, 1983, p. 218). It involves *teaching* such things as data gathering, critical thinking, organizing information, systematic goal setting and self-management. These components are taught as direct curricular content, and the exercise of such skills is reinforced and enhanced through planned practical exercises (Candy, 1991). At present, no research seems to exist that reports on the effectiveness of this *direct* approach.

The second approach to the development of autodidactic competencies is ancillary or concomitant. The philosophy behind this approach is that "autonomous behavior is not taught or learned as ordinary content in the curriculum [...] One learns responsibility and self-direction through experience in which one is given the opportunity to be self-directed and responsible for one's action" (Dittman, 1976). This second perspective has manifested itself in a variety of educational approaches and interventions, ranging from collaborative planning and contract learning to various forms of independent study and *self-directed* learning assignments (e.g., Knowles, 1990; Hiemstra & Sisco, 1990). Several of these interventions have been recently studied. For example, Kasworm (1983) examined the self-directed contract learning as an instructional strategy in a graduate course. She analyzed the impact of a self-directed learning course upon participant self-directed learning behaviors and attitudes. Significant positive gains were noted on participant pre- and postgain scores on the SDLRS. Observational diaries of selected students and of the instructor were analyzed for major themes and transitions. Course evaluations showed a majority of positive participant responses in relation with perceived changes in knowledge and skill in self-directed learning as well as reported value of the course experience.

In a related study, Caffarella and Caffarella (1986) investigated whether using learning contracts in formal graduate education enhanced adults' readiness and competencies for self-directed learning. Their study involved 163 students from six universities in the USA. The students were all enrolled in graduate courses in adult education, where learning contracts were employed. Two testing instruments, the SDLRS and the closely related Self-Directed Learning Competencies Self Appraisal Form (SDLCSAF, developed by the investigators for this study) were administered at the beginning and end of the term by the professors teaching the courses. The findings suggest that the use of learning contracts had little impact on developing perceived readiness for self-directed learning, apparently because perceived readiness was already high among the graduate students attending the classes. The use of learning contracts, however, did have a positive effect on developing perceived competence in self-directed learning, especially in three respects: (i) to translate learning needs into learning objectives in a form that makes possible the accomplishment of these objectives, (ii) to identify human and material resources appropriate to different kinds of learning objectives, and (iii) to select effective strategies for using learning resources.

During the last two decades, learning contracts and various other *ancillary* approaches to the development of autodidactic competences have been used by numerous educators in the USA and abroad. A number of interesting observations have resulted from these experiments. One such observation, due to Biggs (1986, p. 142), is that it is important to provide to learners the opportunity to "talk about their learning processes in a language distinct from that used to talk about the content of their learning." In other words, it is necessary to engage learners in thinking about and discussing their own approach to learning, i.e., engage them in *metalearning* (e.g., Novak & Gowin, 1984), and encourage them to consciously explore alternatives. A second observation concerns the need for scaffolds; learners need to see how, for example, mathematicians think and how they solve problems in real-life contexts, or how economists think about and study particular industries. As Candy (1991) notes, "this may seem self-evident, but there are abundant examples of where an educator has failed to provide any such *scaffolding* within which the learner can erect his or her own pattern of understandings or skill development." A third useful observation is that it is easy in practice to confuse independence and defiance in a learner. As Grow (1991b, p. 218) points out, "A certain kind of student gives the appearance of being a self-directed learner but turns out to be a highly dependent student in a state of defiance. [...] The *false independent* student may resist mastering the necessary details of the subject and try to *wing it* at an abstract level." A last observation concerns the choice between group and individual interventions for the development of self-direction. In a study of eight learners enrolled in a graduate course that attempted to promote self-direction in course work, Taylor (1987) noticed that all students experienced the same sequence of *phases*, which she labeled *disorientation, exploration, reorientation,* and *equilibrium* (Fig. 15-5). Interestingly, she also found that, during the period of 13 wk during which they were interviewed, some learners managed to complete a full cycle and even begin a second one, while others were only engaged in the first two phases: "Learners in this study proceeded at different paces in this cycle." A conclusion one may draw from these observations is that, to accommodate similar situations, educators may have no other option than to individualize their interventions.

Attempts by educators to increase learner-control in formal settings often encounter difficulties. This is hardly surprising in view of the fact that the roles of teachers in this context differ drastically from those to which most educators are accustomed. Some of these difficulties have been analyzed in detail in the literature (e.g., Ainsworth, 1976; Gibbons & Phillips, 1978; Candy, 1991, p. 227-231). They include:

1. a frequent feeling of frustration and helplessness in watching students struggle with problems which the educator knows could easily be solved or avoided (Gibbons & Phillips, 1978),
2. a concern about being *unprofessional* or about being viewed as shirking, abdicating his or her responsibilities as teacher (Harrison, 1978)
3. uneasiness about surrendering a position of authority and superiority,

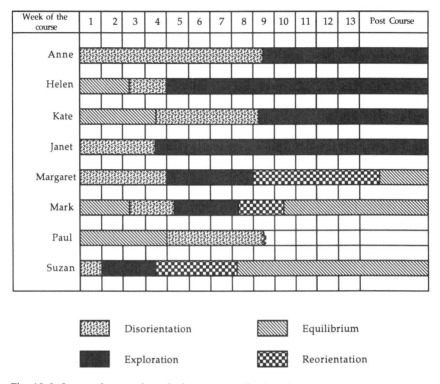

Fig. 15-5. Learners' paces through the sequence disorientation–exploration–reorientation–equilibrium in a course emphasizing self-direction (after Taylor, 1987).

4. the removal of the usual on-the-job reward system of taking credit for student learning (Gibbons & Phillips, 1978),
5. the risk of being regarded as a crank by one's colleagues or by university authorities (Jackson, 1986), and
6. increased rather than lessened demands on the instructor's time.

In this last respect, Ainsworth (1976, p. 279) notes that "there is nothing more effective in the use of the instructor's time than classroom-based instruction, where everything—information, dissemination, test-taking, failure diagnosis—is done according to a schedule, in a group mode, with one explanation serving a large number of students, and where individual assistance is reserved for exceptional cases. Certainly, self-instruction relieves the instructor of the burden of disseminating information, but this is more than offset by the demands of increased individual counselling, and the increased testing, scoring, and diagnosis which commonly accompanies self-instruction."

The difficulty most commonly mentioned by educators is that the students' reaction to participation in self-directed study experiments is often negative. To put it bluntly, students "prefer the conventional method" (Candy, 1991); they "prefer to be taught" (Cornwall, 1988). In contexts where an instructor attempts to promote self-direction, by encouraging students to think about and discuss their own approach to learning, many learners resent and resist these activities as "a waste of time," the instructor avoiding his or her responsibilities," or "not what we have come here to learn" (Baird & Mitchell, 1986). Their initial reactions to these activities often include "shock, confusion and ambivalence" (Taylor, 1987). Many authors have invoked the concept of *learned helplessness* to explain why adults might adopt a passive rather than a proactive attitude toward learning. According to Candy (1987), the argument is that the more people have things done for them, the more *institutionalized* they become, and the more institutionalized they are (in both a figurative and a literal sense), the more dependent, helpless and passive they are. The response of a student, quoted by Eglash (1954, p. 261), vividly illustrates this acquired passivity: "This method [of encouraging self-direction] won't work unless we are brought up in this system and are used to it, and unless everyone co-operates. It allows too much independent thinking."

Even if, as is often argued, years of passivity in formal educational settings deprive many people of the confidence to take charge of their own learning, some educators consider that this tendency can be reverted. As Even (1984) points out, "If such human conditions are learned, they can be unlearned." A firm conviction that such is indeed the case underpins the developments described in the remainder of this chapter.

BRINGING EDUCATIONAL THEORY INTO PRACTICE

General Course Format

In the fall of 1985, I offered for the first time a (biannual) graduate soil physics course specifically designed to foster learner self-direction. The guiding ideas on which the format of this course was based had emerged partially from reacting to my own experience in graduate school, as well as from a few readings, most notably the works of Tough (1967), Knowles (1975), Yourcenar (1980, p. 120), and Brookfield (1983).

These initial guiding ideas were the following:

1. A significant part of the course should consist of some form of individualized *coaching* of the students. I felt, indeed, that individualized coaching sessions, or *tutorials*, would allow me, much better than group discussions or formal lectures, to accommodate the different learning styles and degrees of autonomy of the students. Furthermore, even though knowledge is socially construed and, as a result, much learning occurs informally, or incidentally, in groups (e.g., Marsick & Watkins, 1990; Candy & Crebert, 1991), I considered that individual tutorials would be a much better preparation for the often solitary learning projects that students have to carry out once they leave the university environment.

2. The tutorials should be structured in such a manner as to progressively bring the students to a stage, at the end of the semester, where they

feel reasonably confident they can acquire new knowledge on their own (without an instructor).

3. The course should not consist entirely of tutorials. Indeed, the instructor[1] is himself a learner and is likely to have, or to have had, interesting experiences which would be most efficiently communicated to all the students at once, in a traditional classroom setting. Formal lectures would also make the course appear slightly less eccentric, and thereby perhaps less intimidating to some of the students. In addition, a set of formal lectures would be particularly useful if they could outline a general conceptual framework in which the students could, at a later time, easily situate the materials used for the tutorials. This framework should, however, be presented as the instructor's own rationalization. The students should be strongly encouraged to gradually elaborate their personal reference frame, as well as to formulate their own opinion on some of the controversial aspects of the topic covered by the course.

Since 1985, these three guiding ideas have remained virtually unchanged. They still constitute the basis of the graduate soil physics course in its current format. The practical implication of these ideas has however evolved noticeably over the years, in part because of the reactions of the students themselves. The primary objective of the course has also changed appreciably over the years. In 1985, it was to foster autodidactic learning skills; however, after reading Candy (1990) and some of the other authors referred to in the previous section, I came to realize that a more appropriate goal for this course is to help students to exert increasing control over their learning. As an instructor operating within the context of a course, I have indeed no way at all of influencing the independent learning of the students outside this course or once the semester is over. All I can do is hope that, at some point, they will be able without too much trouble to make the transition from learner-control to autodidaxy.

As currently offered at Cornell, the graduate soil physics course *SCAS 667* deals in detail with the equilibrium physics of aqueous solutions in soils, emphasizing fundamental principles and measurement processes. It serves as an advanced entry into the transport (nonequilibrium) aspects of soil physics and its enrollment generally consists of students who are intent on pursuing a research career in soil physics. The course consists of two components lasting the entire duration of the semester (15 wk): a series of formal lectures and three 5-wk sequences of individual tutorials. These two components are described in detail in the following two sections.

Formal Lectures

Nothing in the format of the lectures distinguishes them from the traditional *top-down* approach. Instructor and students meet three times a week

---

[1] In the following, I shall keep referring to the faculty member as the *instructor*, even though it will become rapidly clear that his or her role is definitely more that of a *helper* or *learning* facilitator (see description, in Rogers, 1969; Vaines, 1974).

for 50 min in a classroom setting, where the students are acquainted with a number of topics selected by the instructor. As in other courses, the instructor tries, whenever possible, to make the lectures lively and attractive by using slides, audiovisual equipment and classroom demonstrations of equipment, or by inviting guest speakers.

Where these lectures begin to depart from tradition is in the selection of the topics they cover. In graduate courses, it is a generally accepted practice for teachers to concentrate their attention on specific, *pet*, subjects that they find particularly interesting or on which they feel comfortable lecturing. Teachers are not required to present a balanced view of a whole field but, rather, consider their role to be that of transferring to the student, at an advanced level, a certain body of knowledge. By contrast, for the formal lectures of the graduate soil physics course, the topics are chosen and structured in such a way as to provide to the students the broadest possible overview of the subject of the course. In the process, depth of coverage is inevitably sacrificed for breadth; however, experience indicates that this approach puts the students in a much better situation when, during the tutorials or after the course ends, they have to acquire new knowledge on their own by reading a scientific article or a book chapter. With a good grasp of only a portion of the field, they would feel in many ways disarmed if the new material they needed to read were not directly related to one of the topics they were previously introduced to. On the other hand, a soil physicist who has been exposed to a broad coverage of the field is likely to find rapidly where the new material fits in the picture and how it is connected to a number of related topics. Confronted with an article on the dielectric properties of soils, for example, he or she would be able immediately to relate its results, e.g., to techniques of soil moisture measurement using time domain reflectometry or ground-penetrating radar, to observations of the structure and dielectric properties of water close to solid surfaces, or to coupled transport processes in the presence of applied electric fields. These connections would be readily apparent to him or her, even if the introduction of the article failed to mention them explicitly.

A convenient analogy may be drawn between what has just been discussed and jigsaw puzzles. Toy manufacturers found long ago that printing a picture of the completed puzzle on the box containing the pieces helps considerably those who try to solve the puzzle. Also, a picture of the whole puzzle, even if it is of somewhat poor quality, is much more useful than a high quality picture of only a small portion of the puzzle. As illustrative as this analogy may be, there is however a key difference. Unlike with puzzles, indeed, the *final picture* or conceptual framework in the graduate soil physics course may not be the same for everyone. The selection and organization of the materials for the formal lectures is based on the instructor's conceptual framework. To avoid imposing his or her framework on the students, and to encourage them to progressively develop their own frame of reference, the instructor should try to make apparent, during the whole semester, the logic of his selection of topics and written sources. He or she should also insist on the fact that conceptual frameworks are personal constructs, varying from individual to individual, in other words that every soil physicist has a different view of what soil physics is about, and of how its various parts are interrelated. In the course, a discussion on this subject is held during the first lecture of the semester, when I briefly review with the students the lecture schedule and the list of reading assignments. At the end of the semester, I raise the issue again during an open discussion.

A second nontraditional aspect of the lectures is the way I handle controversial aspects of the field covered by the course. Some teachers, consciously or unconsciously, seem to consider that graduate students, like undergraduate students, would be utterly confused if they were exposed to some of the controversies that frequently crop up in the literature. From this standpoint, many teachers therefore think it preferable either to entirely avoid mentioning controversial topics or to present only their own views, without dwelling on the fact that other scientists hold markedly different opinions on the same issues. The latter option is undoubtedly the best way to make sure, at least in the short run, that students side with the teacher in some of these controversies; however, in the long run, it may have disastrous effects on the ability of the students to learn autonomously once the course is ended. Indeed, sooner or later, they will encounter articles, or meet scientists, advocating very different viewpoints. If they have never been encouraged or *coached* to see behind the rhetoric and to deal with these controversies in a proactive manner, they may find themselves severely inhibited in their learning process. Therefore, in the formal lectures, I pay particular attention to the areas where researchers have not yet managed to reach a clear consensus. I try, as much as possible, to present to the students a balanced account of the status of the ongoing debates in these areas, using articles from the literature to illustrate the various existing viewpoints. With Frick (1982, p. 197), I believe that the thorough exploration and evaluation of the literature, which this exercise requires, allows the students to "see the chosen discipline as a living entity rather than a static corpus of facts." The students also begin, in the process, to develop judgment.

A corollary of this reliance on recent literature sources is the fact that it becomes difficult for the instructor to assume in this context the traditional role of sage on the stage (Reigeluth & Garfinkle, 1992). As Alain (1908) acidly remarks, some instructors attempt, much like actors, to make their audience believe that they invent as they go the material they present in the lecture room. Without necessarily going to that extreme, many science instructors seem to consider that a key to asserting their authority in the classroom is to conceal their own learning process. Practically, they lecture as if they had always fully mastered the subject of their course; they act as if they never had to experience what their students are going through. The students, of course, know better, but rare are those who dare ask to the *expert* how he or she learned what he or she knows.[2] Fortunately, the situation is

---

[2] There are however exceptions to this general tendency. In the field of the history of higher education, for example, experts and beginners alike are referred to, and refer to each other, as *students of higher education*. The kinship that this creates greatly facilitates the sharing of *tips* on learning.

quite different when the written materials used in class are copies of recently published articles. The more recent the articles, the more blatantly clear it becomes to the students that the instructor is just another student like themselves: it would be futile of him or her to pretend having always known something that only a few weeks or months prior to the lecture was considered original and novel enough to be published in a scientific journal. Obviously, he or she had to do some work on the articles on his or her own in order to comprehend their content. Realizing this, some students feel much freer to ask direct questions to the instructor about his or her learning process, and to try to benefit from his or her experience. This questioning, however, often occurs informally after class or during casual meetings outside the classroom. To allow the largest possible number of students to take advantage of the answers to these questions, it seems preferable for the instructor to be quite explibit about his or her own learning and to volunteer information about it during the lectures. This is the option I have adopted. After I have introduced a particular theory (e.g., fractal theory or the theory or stochastic processes), of which I propose to illustrate applications, I interrupt for a few minutes the natural flow of the lectures and try to explain to the students in detail the process I have used to learn about this theory. More specifically, I describe (and bring to the classroom) the various references and written resources I have consulted during my learning, I explain what led me to these particular references and I try to be honest about the difficulties I may have encountered in obtaining from them the needed information.

The last aspect of the formal lectures that may be worth mentioning is the fact that students are not examined directly on the material presented in the classroom. At the beginning of the semester, the students receive an extensive list of references, as well as a suggested schedule for the readings, but I make clear to them that they should expect no weekly homeworks, no mid-terms and no final exam on the content of the lectures. This is due to the fact that, as was mentioned earlier, one of the primary objectives of the lectures is to encourage the students to develop a conceptual framework of the subset of soil physics dealt with in the course. To check that the students are moving along toward that objective, or at least are establishing a strong scaffolding that in due time will allow them to construct such a framework, homeworks, mid-terms and finals are possible options. However, since in parallel with the formal lectures, I meet with each of the students individually for an hour every week, I have a unique opportunity to assess the student's progress more clearly and more continuously during the semester. In addition, and most importantly, I can actively involve the student in the process, so that this assessment progressively becomes guided or assisted self-assessment on his or her part. This is described in more detail in the following subsection.

Tutorials: Philosophy and Method

In spite of all our efforts to make them nontraditional in a number of ways and, in particular, to make them convey information about the learning process of the instructor, there is no escaping the fact that the formal lectures described above scarcely provide any means or incentive for the students to take charge of their own learning. The instructor no longer presents himself as an *omniscient sage* but he or she is nevertheless still performing in front of the students. The latter are free to, and often do, choose to remain entirely passive if they are so inclined.

For this reason, the formal lectures are not the central component of the graduate soil physics course. They only serve to set the stage for the individual weekly tutorial sessions, in which much of the active learning and most of the metalearning occur. The general objective of these tutorials is to place the students in a situation they will face routinely during their professional career: they are confronted with a scientific article or a book chapter that they have never seen before and that covers an area of direct interest to them. The role of the instructor during the tutorials is to help the students cope with the learning challenge this new material represents for them. The tutorials last ≈ 1 h and take place in a room where the instructor and the student run little risk of interruption.

To avoid the monotony that would result from always working on the same narrow topic for 15 wk, we select three different articles or book chapters of a length typical of journals like the *Soil Science Society of America Journal, Water Resources Research,* or *Soil Science*, and we devote successively five weeks to each of them. Choosing appropriate articles is a critical step. I have found that applied soil physics articles, e.g., on practical aspects of soil tillage or erosion, are in general of little value as a starting point for tutorial sequences. Indeed, except for a few routine statistical techniques, these articles require little prior knowledge, tend to be self-contained and present no serious learning challenge to graduate students. At the other end of the spectrum, some articles (e.g., Sposito, 1982; Maneval et al., 1990) require so much background knowledge in physics or mathematics that it would be difficult for most students to master their content within the time frame of a tutorial sequence. The instructor therefore has to make sure that the articles that are selected within this wide spectrum have a manageable, but sufficient, number of theoretical prerequisites and will take roughly 5 wk for the students to handle. Of course, what may be appropriate for one student, e.g., with a good grasp of physics, may not be so for another.

Ideally, one would want to leave the students entirely free to decide which articles they will use for the tutorials; however, for reasons that were just outlined, it is important for the instructor to retain some control over the selection of these articles. In the last few years, I have tried to strike a compromise, whereby the instructor selects the article for the first tutorial sequence and then gives incrementally increasing freedom to the students to select the second and third articles. By the time the third article has to be chosen (during the ninth week of the semester), the students can generally estimate accurately the difficulties involved in particular scientific articles and they can choose adequate ones with little input from the instructor.

The last time the graduate soil physics course was offered, I assigned to all the students, for the first tutorial sequence, the third section of Babcock's (1963, p. 471–480) seminal paper. This section deals with the theory developed by Gouy (1910) for the distribution of ions in the vicinity of electrostatically-charged planar particles. What makes Babcock's (1963) mathematical description particularly well suited to serve as a basis for a tutorial sequence is the fact that it has a manageable number (6) of implicit but easily identifiable prerequisites. For example, Babcock (1963) assumes that the readers are familiar with the physical meaning of the Poisson and Boltzmann equations, and he makes no references to books or treatises where the assumptions embodied in these equations are analyzed in detail. The students enrolled in the course did not experience any significant problem identifying these and other prerequisites as potential obstacles for a complete understanding of the material covered in the text.

In each of the three tutorial sequences, I encourage the students to proceed step by step through the following *program* or list of activities that results from a careful analysis of my own autodidactic learning projects and from discussions with students over the years.

1. Getting an overview of the article
    Read through from beginning to end and determine the objectives and major results of the article. What is (are) the author(s) trying to do?
    Read through again, but at a deeper level. Determine the aspects of the article that require background reading, i.e., the *stumbling blocks*.
2. Setting goals
    Determine the level at which to read and understand the article.
    Review each of the stumbling blocks identified earlier and set learning goals concerning each one individually.
    Write these goals down in the form of a *contract* for later reference.
3. Elaborating a plan of attack
    For each of the stumbling blocks, think of ways to obtain additional information at a level consistent with the stated goals and locate appropriate sources in libraries. If necessary, seek advice and assistance from colleagues or former professors.
4. Learning
    For each of the stumbling blocks, extract the needed information.
5. Third complete reading and evaluation of learning
    Determine if goals have been met. If necessary, return to step 3 and repeat the process.

The first step in this program is a variation of a method suggested by Descartes (1644) >300 yr ago in the introduction of one of his philosophical works. Descartes' description, somewhat condensed (Houle, 1964) is as follows: "I shall wish the reader at first to go over the whole of the book, without greatly straining his attention, with the view simply of knowing in general the matters of which I treat. Afterwards, if they seem to merit a more careful examination, he may read it a second time, in order to observe the connection of my reasonings, marking the places which he wishes to examine further, and continuing to read without interruption to the end. Then, if he takes up the book a third time, I am confident he will find a solution to most of the difficulties he had previously marked." The effectiveness of Descartes' method has been established repeatedly by psychologists and educational theorists. McClusky (1935), for example, found that when people first skim and then read carefully, their total reading time is less than when they plow straight through the pages, expecting to get all the meaning by a single reading.

In soil physics, unfortunately, it is rarely sufficient to read an article or book chapter a third time to make it suddenly become crystal clear, if it was not already so in earlier readings. Nevertheless, the idea of going through a text in successive waves is one that I have found very valuable in the individual tutorials. Of particular importance is the point where the reader is "marking the places which he wishes to examine further." In the tutorials, I refer to these as *stumbling blocks*. In most cases in soil physics articles, they are caused by reference to mathematical or physical theories with which the reader is not sufficiently familiar. They may also be due to shortcuts taken by the authors in mathematical derivations or proofs of theorems. Finally, they may be related to the principles and limitations of particular measurement methods (e.g., magnetic resonance imaging, neutron scattering, or confocal laser microscopy) with which the reader is not sufficiently acquainted. In all cases, they would represent obstacles for a thorough understanding of the article or book chapter that one is trying to read. Being able to identify these stumbling blocks is crucial to becoming efficient self-directed learners or autodidacts.

In the tutorials, each student is asked to come to the first meeting of each sequence with a list of the stumbling blocks he or she has identified in the article chosen for that sequence. Through discussion with the student, the instructor tries to establish whether the list is complete or whether the student has missed a number of stumbling blocks. If it is conducted tactfully enough, so that the student does not feel submitted to an oral examination, this discussion can provide a lot of information on the ability of the student to diagnose accurately his or her learning needs. This is particularly important during the first tutorial of the semester, where the instructor should try to get a good feel for the level of readiness of students for self-directed learning. If the assessment of this initial level of readiness is not done accurately, there may be a serious mismatch between the role taken initially by the instructor and the learning stage of the students. Some of the problems that may result from such mismatches have been analyzed, e.g., by Candy (1991, p. 410) and Grow (1991a, b).

In preparation for the second tutorial, each student is asked to think carefully about learning goals. Some students are initially puzzled or disconcerted by this assignment. They have for the most part never been acquainted with the idea of formulating clear goals about topics they know little or nothing about, and at first this may appear like a daunting task. In this respect, Candy (1991) notes that "there is a paradox relating to the notion of learners setting their own goals and making reasoned choices from among alternatives." Lawson (1979) goes even further: "What has not yet been learned is not yet known, and the potential learners can only at best dimly perceive what they want to know more about." Fortunately, the situation facing scientists is seldom as gloomy as that and students during the tutorials usually realize quickly that Candy's (1991) paradox is, in their case, only apparent. Indeed the stumbling blocks that they have by then identified in the article or book chapter they are studying provide a list of potential learning objectives.

For each one of these stumbling blocks, each student is confronted with a wide array of alternative choices, ranging from not doing anything at all (i.e., leaving the stumbling block untouched) to consulting an authoritative treatise on the topic. Which option is eventually selected must depend on what the student wants to get out of the article, i.e., at what depth he or she intends to read it. This concept of *learning* (or reading) *depth* has been recognized by many educational theorists (e.g., Houle, 1964; Häyrynen and Häyrynen, 1980; Candy, 1991) who usually distinguish between surface-level and deep-level learning. Surface-level learning of an article like Babcock's (1963) may be adequate for a student who wants to know the rudiments of Gouy's theory in order to, e.g., understand qualitatively how salts affect the hydraulic conductivity of soils. On the other hand, deep-level learning of the same article would be desirable for a student who, e.g., wants to understand the intricacies of the current debate on the mechanisms of clay swelling (e.g., Baveye & McBride, 1994). Obviously, the specific learning goals for each of the stumbling blocks in the first case will be very different than in the second; however, once the student has decided on a targeted depth of learning for the article as a whole, these goals can be set rather easily. Some of them, when they are eventually written down in the form of a *contract*, sound like weather forecasts (e.g., shallow-to-moderate or moderate-to-deep level) but they carry enough information to enable the student to elaborate a strategy or plan of attack (Step 3 in the above program).

This third step also takes place normally during the week preceding the second tutorial. For each of the stumbling blocks, the student is encouraged to locate and obtain information at a level consistent with the corresponding learning goal. In this process, he or she does not have to work in isolation: he or she can seek assistance from colleagues, professors, friends or, last but not least, reference librarians (e.g., Penland & Mathai, 1978, Chapter 2). From these various people, the student may sometimes obtain directly the information needed. For example, if a stumbling block is related to a given shortcut in a mathematical derivation, someone may be patient enough to sit down with the student and solve the problem. Often, however, the best the student can hope to get from conversations with others is pointers to the relevant literature. Whether or not he or she first use human resources, I encourage the student to think carefully about each learning goal before dashing to the library and starting to browse haphazardly through the stacks. Usually, the targeted learning levels will indicate a reasonable starting point in the search. For example, a student seeking a superficial understanding of the Boltzmann equation may want first to consult a freshman physics textbook or a scientific encyclopedia. Contrastedly, a student intending to get a very good grasp of the assumptions embodied in this same equation should probably direct his or her attention immediately to the statistical mechanics literature.

When the learning goals have been clearly formulated and a strategy has been elaborated, the time has come to learn, i.e. extract the information from the selected references and digest it. This process takes place during the following 2 wk in each tutorial sequence. The role of the instructor is to help the students keep their direction, use libraries efficiently, avoid unnecessary aimless wandering in the literature and use appropriate reading styles. This latter point is sometimes a bit obscure to students. Just as there are various possible learning depths, there are different reading styles. Houle (1964) has described a number of them, including "reading to get the central idea", "reading for mastery of content," and "reading to discover a fact or facts". The first two styles are used in Descartes' approach, mentioned above. Students should be encouraged to adopt the third style, which is a kind of *skimming*, whenever they need specific information on a given, narrow topic. This seems like a simple enough thing to do, by using the table of contents or the subject index of the books one wants to consult, without having to read them from A to Z. There are, however, surprising numbers of students who cannot manage it properly, claiming that they have to become thoroughly familiar with the nomenclature and symbols used by an author before they can extract the information they need. Coaching by the instructor is necessary in some cases to overcome this difficulty.

The instructor can also provide valuable help, at least in the first few tutorials, by "encouraging learners to attribute success to their own ability (hence encouraging an optimistic prognosis) but failure to a lack of effort (which the learner can do something about)" (Biggs, 1987). This type of feedback on the students' performance may sound a little paternalistic but one has to remember that, for the most part, the students are mapping what is for them entirely new territory. The risk that they loose faith in their abilities is sometimes very high, at least in the initial phases of the process, and a bit of motivational drill does not hurt.

The last step in the above program takes place during the fifth week of each tutorial sequence. It consists of the ultimate, thorough reading of the article or book chapter being studied in the sequence, and the final self-assessment of learning. This latter part is often by far the most difficult of the whole exercise, largely because the notion that they can evaluate their own learning has been *educated out* of many graduate students: they have been conditioned to think of learning solely as an *achievement-oriented* activity (Candy, 1991) in response to norms and requirements set by others. During the first tutorial sequence, it is frequent for students to come up with a very superficial assessment of their completion of the original *contract*. In these cases, a little probing by the instructor is necessary to make the evaluation more thorough. This has to be done carefully because some students resent this probing and see in it a disguised form of oral examination with no prior notice. The instructor also wants to avoid giving the impression to the students that he or she is trying to impose on them a particular pattern of self-assessment and his or her set of criteria. This would only lead to *achievement-oriented* self-assessment, which would be of little value to the students. By about the tenth week, I have found usually that this probing by the instructor becomes less and less necessary.

In the above description of the tutorials, the role I have envisioned for the instructor is initially a very active one: he or she selects the material to be studied in the first 5-wk sequence and intervenes frequently in the course of the tutorials to put the students back on the right track, if necessary. For the experience to be successful, however, each student should eventually get to the point, at the end of the semester, where he or she feels totally in charge of his or her own learning: the instructor should by then have become a mere *helper*, watching the student from the sidelines and trying to interfere as little as possible with his or her learning. This critical transition from *teacher* to *helper* is illustrated masterfully by Virgil's guidance of Dante through Hell to Heaven in Dante's Divine Comedy (Dante, 1961; Daloz, 1986) and has been studied in detail in the literature on self-directed learning where its importance has been emphasized time and again. Candy (1991), for example, refers to the need for a "progressive devolution of control to the learners." Pratt (1988) talks about a sort of *staged withdrawal* in terms of both support and direction as the learner becomes more accomplished in the domain and more confident of his or her own abilities. My experience in trying to implement such a *staged withdrawal* in the 15 wk timeframe of the graduate course indicates that the pattern of control devolution should be different for each student, in particular because each of them starts off with a different level of self-directedness and, often, very different forms of inhibition.

As presented above, the five-step program for the tutorials appears linear, following closely the pattern described by Knowles (Fig. 15-4). This is in apparent contradiction with the conclusions reached in the review section above, where the largely serendipitous, nonalgorithmic nature of autodidactic projects was highlighted. In fact, this program is followed rigidly only during the first tutorial sequence, where the students already have a large number of new concepts to assimilate and where I feel that it is good for them to acquire a certain learning discipline, i.e., learn an efficient methodology to learn autonomously. In the second and third tutorial sequences, I encourage them to take progressively more and more distance with the above program, while stressing that they should do so only when they feel, after reflection, that there is something to be ultimately gained by it. For example, if a given stumbling block in an article reveals itself much more complicated to resolve than was anticipated during the goal-setting step, the related goal may be somewhat adjusted during the learning phase. On the other hand, the learning step may reveal an aspect of the theory that was neglected or not covered appropriately by the author(s) of the article or book chapter under study. This may suggest another goal that should be added to the now evolving contract.

Difficulties with the Approach

In experimenting since 1985 with the course format described above, I have identified a number of difficulties, none really insurmountable, but of which the instructor has to be keenly aware if he or she wants the course to be successful.

The first of these difficulties is related to the grading of the course. I have found this to be a very difficult issue and I do not feel that it has been resolved satisfactorily. Some institutions of higher education, like the Johns Hopkins University, allow only pass/fail grades in graduate courses. When given the choice, however, students prefer to choose a letter grade option, on the grounds that it looks better on their transcripts. I have found this to be the case for most of the students who took my graduate soil physics course over the years. Even though virtually all of them had a very high motivation level and performed very well, the responsibility for the instructor to decide whether a given student receives a B or a C at the end of the semester is somewhat conflicting with the idea of putting the students in charge of their own learning. In the last couple of years, I have attempted as much as possible to involve the students in the process of awarding the final grade. During the last tutorial of the semester, I ask each student to give an assessment of his or her performance in the course. I then discuss this self-assessment with them and try to come to an agreement. The range of grades, from B to A+, has been so far narrow enough that agreements have rarely been hard to reach; however, this may not always be the case.

The second difficulty, already mentioned earlier, is the high demand on the instructor's time. Compared with a conventional course on the same topic, the proposed format requires on average an additional hour and a half per student per week. This includes the individual tutorial session itself (1 h) and the prior preparation by the instructor. As long as research productivity remains the key criterion for tenure, promotion, and pay raises, this increased time demand will probably be viewed by many as a luxury they cannot afford. I have found that, however, in many cases, the tutorial sessions are a learning experience as much for the instructor as for the student. During the preparation of the tutorials, the instructor indeed has to read articles in more detail than he or she probably would at other times. This makes the instructor more aware of work done by others in his or her field of specialization and in the long run, may make him or her a better researcher. In addition, students often come up with very original ways of looking at particular pieces of research, from which the instructor may also greatly benefit. In cases where time is an insurmountable constraint, it is possible to envisage some modifications of the format of the tutorials. The first sequence (first 5 wk) could for example be deindividualized, since all the students are confronted with the same material. The resulting group-tutorials would then be very similar to the *skill-practice exercises* advocated by Knowles (1990, p. 130); however, because students evolve at different paces (Fig. 15-5), this approach has obvious shortcomings and extending it to all tutorials is not recommended.

Another difficulty with the proposed approach is for the instructor to resist being drawn in as a conventional lecturer during the individual tutorials. Students sometimes look pathetically helpless during their weekly meeting with the instructor and the temptation is then very strong for the latter to assume his or her traditional role and to give a formal lecture to the student. This of course defeats the purpose of the tutorials and has to be avoided as much as possible; it is preferable to spend an hour with the student discussing the problems he or she is experiencing than to provide, predigested, the needed information.

One of the key difficulties with the approach described above is related to the response of the students. Most of them enjoy very much the tutorials, and, in general, the whole structure of the course. For example, one student commented that the course was "the most challenging one he had ever taken." On the other hand, a small number of students dislike the course and feel that it would be much easier for them to learn within a traditional format. One such student, for example, commented that he would have preferred the tutorial sessions to be *real* tutorials (i.e., individualized instruction rather than exercises in self-directon). Every time I taught the course described above, there was a demurring student like this in the class (out of 5 to 7 students). Every one of these students wrote a negative evaluation of the course at the end of the semester. In colleges and universities where evaluations of courses by students are given significant weight in decision processes concerning, e.g., tenure, promotion, or pay raises, very unfavorable evaluations may have serious repercussions for the instructor. Many administrators, fortunately, are far-seeing enough to appreciate the fact that, to be at all meaningful, evaluations of a course like the one described above should really take place 5 to 10 yr after the students have taken the course.

More serious is the risk that, in the long run, students who feel confident only when they are being spoon-fed by a teacher, and who therefore would probably benefit most from the individual tutorials, will shy away from the course and will enroll in alternative offerings with a more traditional format. This risk may be partially alleviated by trying in informal discussions over lunch or coffee to make one's colleagues aware of the autodidactic nature of most of their learning. If they themselves become convinced of the usefulness of a course that attempts to foster autodidactic competence, it is likely that they will strongly advise their graduate students to enroll in the course.

Encouraging students to take charge of their own learning may be a laudable objective; however, when it is pursued in formal institutions where the merits of autodidaxy are not always recognized, there is a definite risk of transforming the students into misfits, ill-adapted to the constraints and requirements of the system. Some of the students who took the graduate course described above and who, by the end of the term, had developed confidence in themselves as independent learners, indeed seemed to have difficulties afterwards to comply with the teachers' requirements in other courses or to take for granted the teachers' viewpoints. They had apparently developed a "taste for more control" (Campbell & Chapman, 1967) of their learning and were reluctant to relinquish this control, even temporarily, to others. One possible way out of this difficulty is to spend some time, during the last tutorial of the semester, discussing with the students about ways to channel their independence in personal learning projects, parallel to the courses they are required to take.

DISCUSSION AND PERSPECTIVES

As every educator knows, each time a course is taught both its format and content evolve. The graduate course described above is no exception. Since 1985, the format of the lectures and, particularly, that of the individual weekly tutorials has changed appreciably. This trend will undoubtedly continue in the future.

A likely innovation that I shall introduce the next time the course is offered (in the spring of 1994) is to encourage the students to read thoroughly the present chapter before the beginning of the semester. Instead of a general introduction to the educational theories of self-direction in learning, the first class meeting of the semester will then consist of an open discussion session. Another direction that I shall pursue is to ask the students to keep a detailed diary of the learning activities they carry out in conjunction with the tutorials. In this diary, they will be encouraged to describe step by step their learning process, to mention explicitly the human or written sources they consult, to write any comment they may have, and to express their frustration if they feel like it. Examples of observational diaries, like those excerpted by Kasworm (1983), will be provided to the students to give them an idea of what is expected of them. These diaries will be used in the guided self-evaluation during the last session of each tutorial sequence and may serve eventually, in the years to come, in the context of a scientific analysis of the tutorial format. A last change that I shall introduce will be to encourage the students during the semester to summarize schematically their own evolving concept of the general framework of the equilibrium physics of soils. A particularly useful tool to achieve this is the so-called *concept-mapping* of Novak and Gowin (1984); boxes serve to visualize the various elements and subcomponents of a given field of knowledge, while arrows indicate the interrelationships between these elements.

A criticism that can be, and probably will be raised against the course format described above is that, as Gruber (1965) puts it forcefully, "there is little reason to believe that a single brief experience with self-directed study in an educational atmosphere fundamentally hostile to intellectual independence will produce attitudinal changes of great longevity." This very sound criticism has already been briefly alluded to in earlier sections. At present, there are unfortunately no quantitative data of any kind that would allow me to answer it directly with regard to the course format described above. Such data could be obtained by comparing students' pre- and postgain scores on the SDLRS or on the OCLI, following Kasworm (1983). This approach would be very easy to carry out; however, its results may not be particularly useful in the sense that gains on either the SDLRS or the OCLI would only indicate increases in *perceived* readiness for self-directed learning, which may or may not be well correlated to *actual* readiness for autodidactic learning. One approach that I have not yet explored and which may help to avoid this shortcoming would be to arrange for independent researchers to ask some of the soil scientists who took the graduate course in the past to keep a detailed diary of some of their current autodidactic learning activities. Analysis of these diaries may yield useful information on whether the course has had any lasting influence on the methodologies used by these autodidacts in their learning endeavors.

A possible variation on the same criticism is that if I am convinced that formal lectures and individual tutorials can have a profound influence on the readiness of soil scientists for autodidactic learning, the approach I propose is *too little, too late*. Indeed, the graduate students who enroll in my graduate soil physics course are usually within a year or two from the end of their program. Efforts to make them become efficient self-directed learners should start much earlier than when they are about to leave the university, and should last much more than just a semester. One way to reply to this comment is to argue that under present conditions in many U.S. universities, soil science graduate students are generally poorly prepared to function autonomously as learners. Therefore any attempt, even feeble and late, to improve the situation is better than not doing anything at all. It is clear, however, that in the long run, the course format described above is not the optimal solution and that the preparation of soil science students for their lifelong learning has to start much earlier than at the very end of their doctoral studies. From this viewpoint, I expect that the new paradigm described in the present chapter will be relatively short-lived and that, in the not too distant future, students will be introduced to the concept and practice of self-directed learning as soon as they get to the university. A pioneering, but so far unique, example of this approach is the undergraduate program of the Hawkesbury Agricultural College in Richmond (New South Wales, Australia) whose curriculum was explicitly designed to promote autonomy in learning (see, e.g., Bawden & Valentine, 1984; Bawden, 1988). Programs of this type may themselves be only transitions towards a more advanced schooling system where traditional *classroom teachers* would have no place and where, from kindergarten onward, education would be inspired by the definition given to it by the Irish poet William Yeats: "not the filling of a vessel, but the lighting of a fire". Until this ideal educational system becomes reality, all or part of the (admittedly imperfect) paradigm described in the present chapter may be of some help in preparing graduate students for the type of learning that awaits them once they leave the university environment.

ACKNOWLEDGMENTS

Sincere gratitude is expressed to J. Baveye and G. André who introduced the author at an early age to the practice of autodidaxy, to Dr. W.A. Jury

(University of California at Riverside) who, unwittingly, caused the author's interest in the literature on autodidactic learning, and to Dr. C.W. Boast (University of Illinois at Urbana-Champaign) for his stimulating support and constructive criticisms during the early stages of the elaboration of the tutorial format. In addition, this chapter has benefited greatly from editorial comments, suggestions and encouragements received from Drs. C.W. Boast, P.C. Candy, T.R. Ellsworth, A. Freeman, B. Gowin, R. Hiemstra, M.S. Knowles, H.B. Long, S. Riha, and A. Tough.

REFERENCES

Adenuga, T. 1991. Demographic and personal factors in predicting self-directedness in learning. p. 93–106. *In* H.B. Long et al. (ed.) Self-directed learning: Consensus and conflict. Univ. of Oklahoma, Norman, OK.

Ainsworth, D. 1976. Self-instruction blues. J. Higher Educ. 47(3):275–287.

Alain. 1908. Propos d'un normand. Tome II. Onzième édition. Gallimard. Paris.

Apps, J.W. 1988. Higher education in a learning society. Jossey-Bass, Publ., San Francisco.

Babcock, K.L. 1963. Theory of the chemical properties of soil colloidal systems at equilibrium. Hilgardia 34(11):417–542.

Baird, J.R., and I.J. Mitchell. 1986. Improving the quality of teaching and learning: An Australian case study—The PEEL Project. PEEL Group, Monash Univ., Melbourne, Australia.

Bauer, B.A. 1985. Self-directed learning in a graduate adult education program. p. 41–49. *In* S.D. Brookfield (ed.) Self-directed learning: From theory to practice (New Directions for Continuing Educ. no. 25). Jossey-Bass Publ., San Francisco.

Baveye, P., and M.B. McBride (ed.) 1994. Clay swelling and expansive soils. Kluwer Scientific Publ., Amsterdam, Netherlands.

Bawden, R.J. 1988. On leadership, change and autonomy. p. 227–241. *In* D. Boud (ed.) Developing student autonomy in learning. Kogan Page Limited, London.

Bawden, R.J., and I. Valentine. 1984. Learning to be a capable systems agriculturalist. Program. Learn. Educ. Technol. 21:273–287.

Berger, N. 1990. A qualitative study of the process of self-directed learning. Ph.D. diss. Division of Education Studies, Virginia Commonwealth Univ., Richmond, VA.

Biggs, J.B. 1987. Student approaches to learning and studying. Australian Council for Educ. Res., Melbourne, Australia.

Bok, D. 1986. Higher learning. Harvard University Press, Cambridge, MA.

Bonham, L.A. 1992. Major learning efforts: Recent research and future directions. p. 48–54. *In* G.J. Confessore and S.J. Confessore (ed.) Guidepost to self-directed learning: Expert commentary on essential concepts. Organization design and development, King of Prussia, PA.

Boud, D. 1988. Moving towards autonomy. p. 17–39. *In* D. Boud (ed.) Developing student autonomy in learning. Kogan Page Limited, London.

Brockett, R.G., and R. Hiemstra. 1985. Bridging the theory-practice gap in self-directed learning. p. 31–40. *In* S.D. Brookfield (ed.) Self-directed learning: From theory to practice (New Directions for Continuing Educ. no. 25), Jossey-Bass Publ., San Francisco.

Brookfield, S.D. 1983. Adult learners, adult education and the community. Open University Press, Milton Keynes, England.

Brookfield, S.D. 1985a. Self-directed learning: A critical review of research. p. 5–16. *In* S.D. Brookfield (ed.) Self-directed learning: From theory to practice (New Directions for Continuing Educ. no. 25), Jossey-Bass Publ., San Francisco.

Brookfield, S.D. 1985b. Self-directed learning: A conceptual and methodological exploration. Stud. Educ. Adults 17(1):19–32.

Brookfield, S.D. 1986. Understanding and facilitating adult learning. Jossey-Bass Publ., San Francisco.

Caffarella, R.S., and E.P. Caffarella. Self-directedness and learning contracts in adult education. Adult Educ. Quart. 36(4):226–234.

Campbell, V.N., and M.A. Chapman. 1967. Learner control vs. program control of instruction. Psychol. Schools 4(1):121-130.

Candy, P.C. 1987. Evolution, revolution or devolution: Increasing learner-control in the instructional setting. p. 159-178. *In* D. Boud and V. Griffin (ed.) Appreciating adults learning: From the learner's perspective. Kogan Page Limited, London.

Candy, P.C. 1989. Constructivism and the study of self-direction in adult learning. Stud. Educ. Adults 21(2):17-38.

Candy, P.C. 1990. The transition from learner-control to autodidaxy: More than meets the eye. p. 9-46. *In* H.B. Long et al. (ed.) Advances in research and practice in self-directed learning. Univ. of Oklahoma, Norman.

Candy, P.C. 1991. Self-direction for lifelong learning: A comprehensive guide to theory and practice. Jossey-Bass Publ., San Francisco.

Candy, P.C., and R.G. Crebert. 1991. Ivory tower to concrete jungle, the difficult transition from the academy to the workplace as learning environments. J. Higher Educ. 62(5):570-592.

Chené, A. 1983. The concept of autonomy in adult education: A philosophical discussion. Adult Educ. Quart. 34(1):38-47.

Chickering, A.W. 1964. Dimensions of independence: The findings of an experiment at Goddard college. J. Higher Educ. 35(1):38-41.

Churchill, W. 1972. My early life. Fontana Books, London.

Collins, M. 1988. Self-directed learning or an emancipatory practice of adult education: Rethinking the role of the adult educator. Paper presented at the 29th Adult Educ. Res. Conf. Univ. of Calgary, Calgary.

Cornwall, M. 1988. Putting it into practice: promoting independent learning in a traditional institution. p. 242-257. *In* D. Boud (ed.) Developing student autonomy in learning. Kogan Page Limited, London.

Craik, G.L. 1830. The pursuit of knowledge under difficulties. Bell & Daldy, London.

Cross, K.P. 1981. Adults as learners: Increasing participation and facilitating learning. Jossey-Bass Publ., San Francisco.

Cross, K.P. 1991. College teaching: what do we know about it? Innovative Higher Educ. 16(1):7-25.

Daloz, L.A. 1986. Effective teaching and mentoring. Jossey-Bass Publ., San Francisco.

Danis, C., and N. Tremblay. 1985a. Critical analysis of adult learning principles from a self-directed learner's perspective. p. 138-143. *In* Proc. of 26th Annual Adult Educ. Conf., Arizona State Univ., Tempe.

Danis, C., and N. Tremblay. 1985b. The self-directed learning experience: Major recurrent tasks to deal with. p. 285-301. *In* Proc. of the 4th Annual Conf. of the Canadian Association for the Study of Adult Education, Montréal.

Danis, C., and N. Tremblay. 1987. Propositions regarding autodidactic learning and their implications for teaching. Lifelong learning: An omnibus of practice and research 10(7):4-7.

Dante (Alighieri). 1961. The divine comedy (J.D. Sinclair, translation). Oxford Univ. Press, New York.

Descartes, R. 1644. Principia philosophiae (English translation by V.R. Miller and R.P. Miller. 1983. Principles of philosophy. D. Reidel Publ. Co., Dordrecht, Holland.)

Dickinson, G. 1979. Teaching adults: A handbook for instructors. General Publ. Co., Don Mills, Ontario, Canada.

Dittman, J.K. 1976. Individual autonomy: The magnificent obsession. Educ. Leadership 33(6):463-467.

Dunleavy, P. 1986. Much to do about knowing. The Times Higher Education Supplement. September 12:15.

Eble, K.E. 1988. The craft of teaching. A guide to mastering the professor's art. (2nd ed.) Jossey-Bass Publ., San Francisco.

Eglash, A. 1954. A group discussion method of teaching psychology. J. Educ. Psychol. 45(5):257-267.

Elton, L.R.B., and D.M. Laurillard. 1979. Trends in research on student learning. Stud. Higher Educ. 4:87-102.

Even, M.J. 1982. Adapting cognitive style theory in practice. Lifelong Learning. The Adult Years 5(5):14-16, 27.

Even, M.J. (chair). 1984. Symposium on adults learning alone. p. 279-284. *In* Proc. of the 25th Annual Adult Educ. Res. Conf., 5-7 Apr. 1984. North Carolina State Univ., Raleigh.

Field, L. 1989. An investigation into the structure, validity, and reliability of Guglielmino's self-directed learning readiness scale. Adult Educ. Quart. 39:125-139.

Frick, E. 1982. Teaching information structure: turning dependent researchers into self-teachers. p. 193-208. *In* C. Oberman and K. Stranch (ed.) Theories of bibliographic education. Bowker, New York.

Gayle, M. 1990. Toward the 21st century. Adult Learn. 1(4):10-14.

Gibbon, E. 1796. Memoirs of my life and writings [Reprinted in 1907 as "Autobiography" Lord Sheffield, (ed.) Oxford Univ. Press, London.]

Gibbons, M., and G. Phillips. 1978. Helping students through the self-education crisis. Phi Delta Kappan 60(4):296-300.

Gouy, G. 1910. Sur la constitution de la charge électrique à la surface d'un électrolyte. Ann. de Physique (Série 4)9:457-468.

Gross, R. 1973. After deschooling, free learning. p. 148-160. *In* A. Gartner et al. (ed.) After deschooling, what? Harper & Row Publ., New York.

Gross, R. 1991. Peak learning. A master course in learning how to learn. Jeremy P. Tarcher, Los Angeles.

Grow, G.O. 1991a. Teaching learners to be self-directed. Adult Educ. Quart. 41(3):125-149.

Grow, G.O. 1991b. The staged self-directed learning model. p. 199-226. *In* H.B. Long and Associates (ed.) Self-directed learning: Concensus and conflict. Univ. of Oklahoma, Norman.

Gruber, H.E. 1965. The future of self-directed study. *In* W.R. Hatch and A.L. Richards (ed.) Approach to independent study. New Dimensions in Higher Education no. 13. U.S. Dep. of Health, Education, and Welfare, Washington, DC.

Guglielmino, L.M. 1977. Development of the Self-Directed Learning Readiness Scale. Ph.D. diss. Univ. of Georgia, Athens.

Guglielmino, P.J., L.M. Guglielmino, and H.B. Long. 1987. Self-directed learning readiness and performance in the workplace: Implications for business, industry, and higher education. Higher Educ. 16:303-317.

Halterman, W.J. 1983. The complete guide to nontraditional education. Facts on File, New York.

Harrison, R. 1978. How to design and conduct self-directed learning experiences. Group Organ. Stud. 3(2):149-167.

Häyrynen, Y.-P., and S.-L. Häyrynen. 1980. Aesthetic activity and cognitive learning: creativity and orientation of thinking in new problem situations. Adult Educ. in Finland 17(3):5-16.

Hiemstra, R., and B. Sisco. 1990. Individualizing instruction. Jossey-Bass Publ., San Francisco.

Houle, C.O. 1961. The inquiring mind: A study fo the adult who continues to learn. Univ. of Wisconsin Press, Madison.

Houle, C.O. 1964. Continuing your education. McGraw-Hill Book Company, New York.

Jackson, M. 1986. Madness in the method. The lecture is obsolete—and students deserve a new approach. Times Higher Educ. Suppl. 14 Nov:14.

Jones, A.N., and C.L. Cooper. 1980. Combating managerial obsolescence. Philip Allan Publ. Limited, Oxford, England.

Kahnweiler, W.M. 1991. Professor-student relationships: Nurturing autonomy or dependency. J. Professional Stud. Spring/Summer:32-41.

Kasworm, C.E. 1983. An examination of self-directed contract learning as an instructional strategy. Innovative Higher Educ. 8(1):45-54.

Kidd, J.R. 1973. How adults learn (rev. ed.) Association Press, New York.

Knowles, M.S. 1975. Self-directed learning: A guide for learners and teachers. Association Press, New York.

Knowles, M.S. 1984. Andragogy in action. Jossey-Bass Publ., San Francisco.

Knowled, M.S. 1990. Fostering competence in self-directed learning. p. 123-136. *In* R.M. Smith and et al. (ed.) Learning to learn across the life span. Jossey-Bass Publ., San Francisco.

Landers, K. 1989. The Oddi continuing learning inventory: An alternate measure of self-directio nin learning. Unpublished doctoral dissertation. Syracuse Univ., Syracuse, New York (Diss. Abstr. ACC9008433).

Lawson, K.H. 1979. Avoiding the ethical issues. p. 123-134. *In* K.H. larson (ed.) Philosophical concepts and values in adult education Rev. Ed. Open Univ. Press, Milton Keynes, England.

Long, H.B. 1991a. Self-directed learning: consensus and conflict. p. 1-9. *In* H.B. Long et al. Associates (ed.) Self-directed learning: Consensus and conflict. Univ. of Oklahoma, Norman.

Long, H.B. 1991b. College students' self-directed learning readiness and educational achievement. p. 107-122. *In* H.B. Long and et al. (ed.) Self-directed learning: Consensus and conflict. Univ. of Oklahoma, Norman.

Low, P.F. 1970. Graduate instruction in soil chemsitry. p. 9-14. *In* H.S. Jacobs and A.L. Page (ed.) Graduate instruction in soil science. ASA Spec. Publ. 17. ASA, CSSA, and SSSA, Madison, WI.

McClusky, H.Y. 1935. An experiment on the influence of preliminary skimming on reading. J. Educ. Psychol. 25:521-529.

Maneval, J.E., M.J. McCarthy, and S. Whitaker. 1990. Use of nuclear magnetic resonance as an experimental probe in multiphase systems: Determination of the instrument weight function for measurements of liquid-phase volume fractions. Water Resour. Res. 26(11):2807-2816.

Marsick, V.J., and K.E. Watkins. 1990. Informal and incidental learning in the workplace. Routledge, London.

Martin, R.A. 1991. Avoiding the catastrophe of human obsolescence. Agric. Educ. Magazine 63(10):4, 10.

Merriam, S.B., and R.S. Caffarella. 1991. Learning in adulthood. A comprehensive guide. Jossey-Bass Publ., San Francisco.

Moore, M.G. 1973. Toward a theory of independent learning and teaching. J. Higher Educ. 44(12):661-679.

Naisbitt, J., and P. Aburdene, 1985. Re-inventing the corporation. Warner Books, New York.

Nielsen, D.R. 1970. Graduate instruction in soil physics. p. 1-5. *In* H.S. Jacobs and A.L. Page (ed.) Graduate instruction in soil science. ASA Special Publ. 17. ASA, CSSA, and SSSA, Madison, WI.

Novak, J.D., and D.B. Gowin. 1984. Learning how to learn. Cambridge Univ. Press, Cambridge, England.

Oddi, L.F. 1985. Development and validation of an instrument to identify self-directed continuing learners. p. 230-235. *In* Proc. of the 26th annual Adult Education Research Conference, Arizona State Univ., Tempe.

Oddi, L.F. 1986. Development and validation of an instrument to identify self-directed continuing learners. Adult Educ. Quart. 36:97-107.

Penland, P. 1978. Individual self-planned learning. Office of Education, Washington, DC.

Pratt, D.D. 1984. Adragogical assumptions: Some counter-intuitive logic. *In* Proc. of the Adult Education Research Conference 25. North Carolina State Univ., Raleigh.

Pratt, D.D. 1988. Andragogy as a relational construct. Adult Educ. Quart 38(3):160-181.

Raven, M. 1992. Teaching students with different learning styles. Agric. Educ. Mag. 65(3)5-6, 15.

Reetz, H.F., Jr. 1972. The learning environment—A graduate student's view. J. Agron. Educ. 1:55-57.

Reigeluth, C.M., and R.J. Garfinkle. 1992. Envisioning a new system of education. Educ. Technol. 32(11):17-23.

Resnick, L.B. 1987. Learning in school and out. Educ. Res. 16(9):13-20.

Rogers, C. 1969. Freedom to learn. Charles E. Merrill Publ. Co., Columbus, OH.

Rohfeld, R.W. 1991. Adult students, adult education and the Ed. D.: An alternative residency experience. Innovative Higher Educ. 16(1):49-58.

Six, J.E. 1987. Measuring the performance properties of the Oddi continuing learning inventory. Ph.D. diss. Syracuse Univ., Syracuse, NY.

Skager, R.W. 1979. Self-directed learning and schooling: Identifying pertinent theories and illustrative research. Int. Rev. Educ. 25:517-543.

Smith, R.M. (ed.) 1990. Learning to learn across the life span. Jossey-Bass Publ., San Francisco.

Spear, G.E. 1988. Beyond the organizing circumstance: A search for methodology for the study of self-directed learning. *In* H.B. Long et al. (ed.) Self-directed learning: Application and theory. Dep. of Adult Educ., Univ. of Georgia, Athens.

Spear, G.E., and D.W. Mocker. 1984. The organizing circumstance: Environmental determinante in self-directed learning. Adult Educ. Quart. 35:1-10.

Sposito, G. 1982. Theory of quasielastic neutron scattering by water in heterogeneous systems. Mol. Physics 47(6):1377-1389.

Taylor, M. 1987. Self-directed learning: More than meets the observer's eye. p. 179-196. *In* D. Boud and V. Griffin (ed.) Appreciating adults learning: From the learner's perspective. Kogan Page Limited, London.

Thomas, A.M. 1967. Studentship and membership: A study of roles in learning. J. Educ. Thought 1(1):65-76.

Tough, A. 1967. Learning without a teacher. Educational Research Series no. 3, The Ontario Inst. for Stud. in Educ., Toronto, Canada.

Tough, A. 1968. Why adults learn: A study of the major reasons for beginning and continuing a learning project. The Ontario Inst. for Stud. in Educ., Toronto, Canada.

Tough, A. 1978. The adult's learning projects: a fresh approach to theory and practice in adult learning. 2nd ed. The Ontario Inst. for Stud. in Educ., Toronto, Canada.

Tough, A. 1979. The adult's learning projects: A fresh approach to theory and practice in adult learning. 2nd ed. The Ontario Inst. fr Stud. in Educ., Toronto, Canada.

Tremblay, N.A. 1981. L'aide à l'apprentissage en situation d'autodidaxie. Ph.D. diss., Université de Montréal, Canada.

Tremblay, N.A., and J.P. Theil. 1991. A conceptual model of autodidactism. p. 28-51. In H.B. Long et al. (ed.) Self-directed learning: Consensus and conflict. Univ. of Oklahoma, Norman.

Vaines, E. 1974. Student-centered teaching. p. 161-172. In E.F. Sheffield (ed.) Teaching in the universities, No one way. McGill-Queen's Univ. Press, Montreal.

Wang, M.C. 1983. Development and consequences of students' sense of personal control. p. 185-197. In J.M. Levine and M.C. Wang (ed.) Teacher and student perceptions: Implications for learning. Erlbaum, Hillsdale, New Jersey.

West, R.F., and E.L. Bentley, Jr. 1991. Relationships between scores on the self-directed learning readiness scale, ODDI continuing learning inventory and participation in continuing professional education. p. 71-92. In H.B. Long et al. (ed.) Self-directed learning: Consensus and conflict. Univ. of Oklahoma, Norman.

Witkin, H.A., C.A. Moore, D.R. Goodenough, and P.W. Cox. 1977. Field-dependent and field independent cognitive styles and their educational implications. Rev. Educ. Res. 47(1):1-64.

Yourcenar, M. 1980. Les yeux ouverts (entretiens avec Matthieu Galey). Le Livre de Poche, Paris.